INDICATORS RELEVANT TO FARM ANIMAL WELFARE

CURRENT TOPICS IN VETERINARY MEDICINE AND ANIMAL SCIENCE

Distributors

for the United States and Canada: Kluwer Boston, Inc., 190 Old Derby Street, Hingham, MA 02043, USA
for all other countries: Kluwer Academic Publishers Group, Distribution Center, P.O.Box 322, 3300 AH Dordrecht, The Netherlands

Library of Congress Cataloging in Publication Data

Main entry under title:

Indicators relevant to farm animal welfare.

(Current topics in veterinary medicine and animal science ; v. 23)
"Sponsored by the Commission of the European Communities, Directorate-General for Agriculture, Coordination of Agricultural Research."
1. Livestock--Congresses. 2. Veterinary medicine--Congresses. 3. Animals, Treatment of--Congresses. I. Smidt, Diedrich. II. Commission of the European Communities. Coordination of Agricultural Research. III. Title: Farm animal welfare. IV. Series.
SF5.I44 1983   636.08'3   83-12178
ISBN-13:978-94-009-6740-3   e-ISBN-13:978-94-009-6738-0
DOI: 10.1007/978-94-009-6738-0

ISBN-13:978-94-009-6740-3
EUR 8514 EN

Book information

Publication arranged by: Commission of the European Communities, Directorate-General Information Market and Innovation, Luxembourg
Proceedings prepared by: Janssen Services, 33a High Street, Chislehurst, Kent BR7 5AE, UK

Copyright/legal notice

INDICATORS RELEVANT TO FARM ANIMAL WELFARE

A Seminar in the CEC Programme of Coordination of Research
on Animal Welfare, organized by Dr. D. Smidt, and held in
Mariensee, 9—10 November 1982

Sponsored by the Commission of the European Communities,
Directorate-General for Agriculture, Coordination of Agricultural Research

Edited by

D. Smidt
Institut für Tierzucht und Tierverhalten der FAL
Mariensee
Federal Republic of Germany

1983 **MARTINUS NIJHOFF PUBLISHERS**
a member of the KLUWER ACADEMIC PUBLISHERS GROUP
BOSTON / THE HAGUE / DORDRECHT / LANCASTER

for

THE COMMISSION OF THE EUROPEAN COMMUNITIES

CONTENTS

P R E F A C E

The CEC-seminar on "Indicators relevant to Animal Welfare" was held on November 9 - 10, 1982 at the Institut für Tierzucht und Tierverhalten, Mariensee, of the Bundesforschungsanstalt für Landwirtschaft (FAL). It was the intention of this meeting to outline the present state of knowledge on proven and potential indicators and to discuss their applicability in the assessment of farm animal management systems. After an introduction of the legal and material aspects of animal welfare, the following topics were dealt with :

- Significance of indicators relevant to animal welfare
- Integrated systems of indicators relevant to animal welfare
- Applicability of indicators in animal welfare research and control procedures.

The concluding discussion focussed on the applicability of indicators in science and in practice.

As a result of the seminar, the presently known indicators can be grouped into the following categories :

- indicators already being applied to animal welfare assessment in practice, such as morbidity and mortality, technopathies, behavioural aberrations and production performances;

- indicators of high sensitivity, such as physiological criteria, which as yet are too demanding for general application;

- indicators still under investigation for their reliability.

It became also evident that the continuous integration of indicators into comprehensive assessment systems promises an ever improving judgement of animal welfare problems.

The proceedings will be of substantial value to the CEC in the pursuit of animal welfare problems.

The CEC gratefully acknowledges the contributions of the participants and the efforts of the organizers. The seminar helped to further promote interdisciplinary research in the area of animal welfare and the cooperation of the European Communities.

SESSION I

GENERAL OUTLINE

Chairman: W. Sybesma

ANIMAL WELFARE LEGISLATION

Prof. Dr. Amin Rojahn, Ministerialrat,
Hans-Dieter Liebich, Oberregierungsrat

Federal Ministry of Food, Agriculture
and Forestry
Rochusstr. 1, 5300 Bonn
Federal Republic of Germany

ABSTRACT

The legal background of animal welfare consists of the
area of legislation and the area of jurisprudence. The state
of the art is differing widely among the various member countries of the EC. The German Animal Protection Act provides for
individual, direct, ethical protection of animals, legislation
in other countries of the EC gives in most cases priority to
the responsibility of man regarding the well-being or welfare
of animals. But one should aim at harmonizing legislation
within the EC - also for the purpose, among other things, of
avoiding distortion of competition. Up to now only very moderate steps have been taken in that direction. In order to make
progress scientists are called upon to define parameters enabling us to assess the keeping of and the care of animals
under the aspects of animal welfare. Divergence of opinions
on the part of researchers presents an obstacle for the enforcement of the necessary legislation.

I. LEGAL BACKGROUND

There are two main areas that constitute the legal background of animal welfare:

- the statutory area; the existing legislation pursuing
  certain purposes and aims, and

- the area of jurisprudence; the administration of justice
  by courts and the discussions on legal aspects of animal
  welfare.

Both areas are closely related - they even condition
each other. But there is an essential difference between them:
Legislation is written, established law. In democratic
countries legislation is the result of the formulation of the
objectives of society. It is the concretion of the sense of
justice that, at the time of the elaboration of a piece of
legislation, is felt to be right and just by the majority of
the members of the national community. Legislation is static.
In the administration of justice and the discussion in the

public, however, the progress of formulating laws, the
dynamics of law-making is reflected.

The process of the formulation of laws is by no means a
matter of mere cognition as it is the case in natural science
processed but is moulded by the interests of certain groups
of society and the will of the public. Thus law is subject to
evolutions based on the development of the moral and ethical
rules of society. Such an evolution can, however, lead to
tensions between the static, established legislation and what
is considered to be just by the public. It is the task of the
administration to take account of this and if necessary seek
a balance. The dynamics of the formulation of laws due to
further developments in the public sense of justice is differing widely among peoples - even in Europe. In the context
of animal welfare this means that in the individual contries
of the EC there is a different level of willingness to improve
animal welfare in a joint effort to harmonize legislation in
this area. In the negotiations in Brussels for example on
keeping laying hens in cages, this has become particularly
evident.

II. THE LEGAL FRAMEWORK

a) The legal basis for animal welfare in the Federal
Republic of Germany is the German Animal Protection Act of
24 July 1972. The basic principles of this Act are Article 1
and Article 2, para. 1 which read as follows:

Art. 1: "This Act shall serve to protect life and well-
being of the animal. Without reasonable cause nobody shall
cause pain, suffering or injury to an animal."

This basic principle contains the commitment on the part
of the legislator to animal protection for ethical reasons.
It aims at protecting animals against man - individually and
directly. It protects the life and the well-being of animals.
Its life is to be protected against destruction, its well-
being against certain restrictions specified in the Act or
against the risk of such restrictions. Simultaneously it protects the freedom from injury of the animal by protecting it

against damage. Health, freedom and dignity are protected
only indirectly if an action directed against them is at the
same time an attack against the life, the well-being or the
freedom from injury of an animal in a prohibited way. As
there are,however, in addition to the values mentioned in
the German Animal Protection Act, other values to which society attaches great importance, the legislator, in order to
find a balance between competing values, created the criterion of "reasonable causes", i.e. in exceptional cases for
sensible, justified and unavoidable reasons it is possible to
subject an animal to pain, suffering or damage. Such sensible
reasons are interests that are given priority over animal
welfare considerations, for example production of food, disease control and scientific research.

The obligations of persons keeping animals are laid down
in para. 1 of Art. 2 of the German Animal Protection Act
which stipulates: "Any person who is keeping an animal or who
is looking after it,

1. shall give the animal adequate food and care suitable
for its species; and he shall provide accommodation which
takes account of its natural behaviour;

2. shall not permanently and not so restrict the needs of
an animal of that species for movement and exercise that the
animal is exposed to avoidable pain, suffering or injuries."

Similar legislation including basic principles is to be
found in other countries of the EC, too.

The Danish Act No. 256 of 27 May 1950 contains the
following principles on the protection of animals:

Art. 1 "Animals shall be treated in a responsible way
and may not be exposed to unnecessary sufferings due to neglect, strain or other causes.

Art. 2 "Any person who is keeping animals shall see to
it that the animal shall get sufficient and suitable food and
water and that it is kept and looked after in a responsible
way, for example by acceptable housing."

Although the ethical aspects of these principles are not
stressed as much, this Act, too, underlines the responsibility of man towards the animal. The animal is considered as a

living sensitive being, as a fellow creature the respect and appreciation of which is a moral postulate for man who is superior because of his intellect. But there are no commitments referring to the principle that animals must be kept in accordance with behavioural needs or other needs of a species or of an animal.

In France Decree No. 80 - 791 of 1 October 1980 on the enforcement of Art. 276 of the Agricultural Land Act,Title 1, Art. 1 also stressed the role of man as the general guarantor of the well-being of animals in his care. According to this provision it is prohibited to keep from animals food and water required by their physiological needs or to let animals in case of illness or injury without care as well as to house animals in a way that may lead to sufferings, injuries or accidents.

Treatment of animals according to their needs, care for the well-being of the animals, prohibition of unjustified infliction of pain, suffering or damage: these are the principles of the Swiss Federal Act on Animal Welfare of 19 March 1978 - principles that are similar to those contained in the German Animal Protection Act but giving priority to the responsibility of man, too. According to the Swiss legislation (Art. 1 of the Swiss Animal Welfare Regulation of 27 May 1981) proper keeping of animals is defined as follows:

1. Animals shall be kept in such a way that their body functions and their behaviour are not upset and that there are no undue demands on their adaptability.

2. Feeding, care and housing are appropriate if, in accordance with the state of experience and the findings of physiology, ethology and hygiene, they meet the needs of the animals.

3. Animals shall not be kept tied in a permanent way.

4. Deviations from provisions on the keeping of animals may be permitted in exceptional cases if they are necessary to prevent or cure diseases.

The point of departure of these demands is not "their being in accordance with the needs of the species", but the reference to the needs of the individual, that is their accordance with the needs of the individual animal concerned.

b) Animal welfare costs money, this is why provisions on animal welfare must be studied in the light of equal opportunities for competition. If a whole industry - like for example agriculture - within one uniform economic zone is exposed to differing conditions for competition this is bound to result in the desire for a harmonization of the legal standards applicable to that economic zone. Here one can strive for the ideal situation but will hardly ever achieve it. In the light of a variety of different opinions and interests the basis to be looked for will in most cases be the lowest common denominator.

In the European Communities, for example, over the last few years a series of directives have been elaborated aiming at a harmonization of animal welfare legislation in the EC. These are: Council Directive 77/489 (EEC) of 18 July 1977 (on the protection of animals during international transport), Council Directive 81/389 (EEC) of 12 May 1981 (measures for the enforcement of the transport directive), Council Directive 74/577 (EEC) of 18 November 1974 (on the stunning of animals before slaughter) and, above all, Decision 78/923 (EEC) of the Council dated 19 June 1978 on the adoption (that means into EC legislation) of the European Convention on the Protection of Animals Kept for Farming Purposes (this convention was elaborated in the Council of Europe in order to protect animals kept on farms especially those kept in modern intensive management systems). In reasons given by the Council for this decision it is stated that animal welfare per se is not among the objectives of the Community.

There were, however, differences between the existing national legislations on the protection of farm animals resulting in a distortion of competition which thus had a direct impact on the functioning of the Common Market. This convention was, therefore, taken over for purely economic reasons. Similar reasons were stated for the other above-mentioned directives on the protection of animals.

In the convention adopted by the Community there are basic principles of modern animal welfare, too. Animals have

8

to be housed and provided with food, water and care in a manner which - having regard to their species and to their degree
of development, adaptation and domestication - is appropriate
to their physiological and ethological needs in accordance
with established experience and scientific knowledge (Art. 3)
and the freedom of movement appropriate to an animal having
regard to its species and in accordance with established experience and scientific knowledge may not be restricted in
such a manner as to cause it unnecessary suffering or injury
(Art. 4).

III. THE AREA OF JURISPRUDENCE

The discussion on the legal aspects of animal welfare and
the court rulings in this area often focus - at least in the
Federal Republic of Germany - on the "interpretation of statement of the facts" of individual pieces of legislation. Obviously there are great uncertainties in this area. These uncertainties, however, are not caused by the lacking ability
of legal experts, to give a clear, definite statement of the
facts. The reason is that natural science experts have hitherto not been in a position to supply legal experts with uniform and generally agreed findings, for example on the keeping
of animals.If the legal expert has to formulate a statement
of facts or to interpret such a statement he is confronted,
in the present state of the art, with major problems.

A statutory statement of facts has the purpose of giving
a generally valid description of a certain situation and to
relate this to a regulation. In a simple case this is simple,
for example if a cow is to be transported in the lactation
phase. People dealing with agricultural matters know, that
lactating cows must be milked at certain intervals to avoid
pain and damage. The regulation for this case, therefore,
reads: Cows in milk shall be milked at intervals of not more
than 12 hours (Council Directive 77/489 (EEC) of 18 July 1977
(protection of animals during international transport) Annex,
Chapter I, Item 9b). This is a clear and precise provision on
how to act in a certain case.

It is, however, possible that a cow or any other animal

does become ill or injured during transport because of unpredictable reasons. How can this much more intricate situation be regulated so that it is at the same time applicable to legal practice? The above-mentioned Directive on transport has stated the facts of the case concerned in the following way: Animals which become ill or injured during transport shall receive veterinary attention as soon as possible and if necessary be slaughtered in a way which avoids unnecessary suffering. (Annex to the above-mentioned Directive, Chapter I, Item 10).

In this context a relatively simple situation is covered by a broad statutory definition which does not make a definite statement on how to do what. It leaves many questions open. On the other hand this is the only way to regulate the protection during transport of animals of most different development levels and with the most different diseases or injuries in a relatively practicable way. If the statutory definition were not so broad there would have to be a regulation on a case-to-case basis which in practice is virtually impossible. If one tries now to regulate animal husbandry in a generally valid way one is bound to come to broad and comprehensive terms like "pain, suffering or damage, appropriate, in accordance with the needs of a species, according to its normal behaviour".

To give a meaning to the vague legal terms in the fields of animal welfare legislation in a concrete case is a major problem to those who are responsible for this area in the Federal Republic and very likely in other EC countries as well; for example the points of view on whether the terms "in accordance with the needs of the animal" or "in accordance with the needs of the species" are better suited, differ widely. There are those who think that "in accordance with the needs of the species" is misleading since in many species there can be subspecies which can be adapted to very different environmental conditions, particularly because of early experience and imprinting processes. But there are others who consider the term "in accordance with the needs of the species" to be the general term including subspecies, strains,

wild or bred forms of one species, providing reliable information on the characteristics of the species concerned which
include a certain range of variants. The third approach is
that it is not a problem of species but of subspecies, populations etc. and that the only suitable term would be "typical
of a species". The requirement in this case must be that an
animal must be free of injury in the normal range of the
species' behaviour, this would be in accordance with the needs
of the animal. The essential needs of an animal must be met;
this could be done by technical means, too, and the animals
could be at least exposed to the level of strain they had to
bear in the wild.

In the Federal Republic, for example, the competent authority is entitled, if necessary, to take measures providing
the animal with food, care and housing that is in accordance
with its behavioural needs or avoiding that the need for movement of the species is restricted permanently or in a way
that the animal is exposed to avoidable pain, suffering or
damage (see para. 2 of Art. 2 together with para. 1 of the
German Animal Protection Act). Before taking its decision the
authority must know when a certain management system is in
accordance with the behavioural needs of the animal concerned,
what the need for movement of a species concerned is, at what
time a certain animal sufferns. Only on the basis of such
knowledge it is in a position to make comparisons and take
its decision. The authority does not have any scope for form-
ing an opinion of its own or to make an assessment of whether
the animal is kept in accordance with its behavioural needs.
It has, in taking a decision, to keep in mind all parameters
a management system must contain in order to be objectively
in accordance with the behavioural needs. It must be possible
for a court of justice to review this decision. The court itself independently gathers the parameters and assesses the
situation in the light of these parameters. If the authority
were entitled to make a subjective assessment it would not be
possible for a court of justice to review this decision.

In legal practice this means: If animal welfare legislation contains vague legal terms as characteristics for a

statement of the facts, these must preferably be terms that on principle can be described objectively and do not allow for subjective assessments and, as a result, for scope of discretion. They can be reviewed by the courts that means the courts must decide on whether a certain management system is in accordance with the recognized conditions of proper management or whether these conditions are not complied with and there is a violation of animal welfare legislation.

The courts of justice in the Federal Republic have had various opportunities - mainly in the context of laying hens kept in cages - to review concrete facts on the basis of existing animal welfare legislation.

In their rulings the courts of justice refer to the problems concerning the statement of facts in animal welfare legislation in practice and compare the differing views of scientists. Because of these differing views the courts have come to different rulings, too. But hitherto there has been no conviction in a case of keeping laying hens in cages. It is true that in the majority of cases courts confirm that in normal cage systems for laying hens the animals are exposed to considerable long-term sufferings without reasonable cause. (District Court of Düsseldorf, 23 November 1979 - II - 23/79; Local Court of Leverkusen, 24 April 1979 - 17 Ls 5 Js 120/77 (178/78) E; Regional Court of Frankfurt, 12 April 1979 - 4 Ws 22/79); but it is stated that the persons concerned cannot be reproached with an offense as they did not act with intent. The reason for the lack of reproachability stated by the courts is "that scientists with an established reputation continue to have arguments on whether such suffering can really be objectified". (Local Court of Leverkusen, 24 April 1979, 17 Ls 5 Js 120/77 (178/78/E-). In other words: If even scientists cannot say when and how an animal is suffering, how should laymen be able to formulate an opinion on the conditions under which their behaviour towards an animal leads to suffering for the animal or at what level this suffering is "considerable".

Judgements of this kind are used for demanding a general ban on certain animal management systems and, in addition to

the Animal Protection Act, further legislation for the keeping
of animals. Such demands are likely to be put forward in some
years' time in other countries of the EC, starting with the
keeping of laying hens in battery cages. It is the task of
the scientific sector to define parameters allowing for an
assessment of management systems that are in accordance with
the needs of the species or the animal, and housing systems
that are in accordance with the behavioural needs of the animals, which at the same time meet the need for movement of
the animals concerned. It must be possible for those who have
to deal with animal husbandry, to see when, in what systems
and in what way the essential needs of the animals are met.
This is the precondition for practicable, enforceable and
verifiable statutory regulations. There is a need for such
regulations, in the first place, on the keeping of calves,
pigs and laying hens which meet the ethological, hygienic and
economic requirements in a satisfactory way.

THE PROBLEM OF ASSESSING "WELL-BEING" AND "SUFFERING" IN FARM ANIMALS

I.J.H. Duncan*, M.S. Dawkins**

*ARC Poultry Research Centre,
Roslin
Midlothian EH25 9PS, U.K.
**Department of Zoology,
University of Oxford,
South Parks Road,
Oxford OX1 3PS, U.K.

ABSTRACT

The terms "well-being" and "suffering" are discussed and it is suggested that it will be very difficult, if not impossible to give them precise definitions. The classes of evidence used as indicators of well-being or of suffering are considered in turn. Indicators associated with health are straightforward; any reduction in health, by definition, means a reduction in well-being. The conclusion about productivity as an indicator is also simple; although theoretically it may have merit, in practice it is a dangerous criterion to use. The main problem associated with using physiological and biochemical changes as indicators is in deciding how much change an animal can tolerate without suffering. Similar problems exist with behavioural indicators; how much of a particular behaviour must an animal show to indicate suffering? Abnormal behaviour may be an indicator of suffering but how do we define "abnormal"?

However, the whole welfare debate hinges on the controversial questions, "Do animals have subjective feelings, and if they do, can we find indicators which reveal them?" The evidence suggests that all agricultural species probably do have feelings although these might be very different from human feelings. There are also indications that with careful experimentation we may be able to accumulate indirect evidence about animals' subjective feelings. This should be our ultimate aim. There are many problems but they are not insurmountable.

INTRODUCTION

It has often been stated that in making the best reasonable estimate of either animal well-being or animal suffering, we should take account of all the available evidence. This will include evidence of the animals' health, productivity, physiology, biochemistry and behaviour (Dawkins, 1980; Duncan, 1978a, 1981). This paper discusses some of the problems associated with questions like, "Which class of evidence should be used?", "Which methods give the most reliable indicators?" and "How should conflicting evidence be resolved?"

The first and perhaps the biggest problem is one of definition. What do we mean by well-being or welfare or suffering? These are common words of everyday speech and, as such, they are used rather loosely and tend to mean different things to different groups of people. This immediately creates a problem for the scientist who wishes to be very precise and give these phenomena exact and unambiguous definitions. We think that this is impossible; people will always read different shades of meaning into terms like "well-being" and "suffering". Nevertheless, it is possible to have some general working definition which broadly describes the area to be covered. In this paper "well-being" and "welfare" are used synonomously and some definitions or, more accurately, broad descriptions of welfare follow.

The Brambell Committee, which was formed by the British Government to investigate the welfare of intensively housed livestock, stated that, "Welfare is a wide term that embraces both the physical and the mental well-being of the animal. Any attempt to evaluate welfare, therefore, must take into account the scientific evidence available concerning the feelings of animals that can be derived from their structure and function and also from their behaviour" (Command Paper 2836, 1965). Hughes (1976) defined welfare as "A state of complete mental and physical health, where the animal is in harmony with its environment". More recently, Carpenter (1980) has said that "The welfare of managed animals relates to the degree to which they can adapt without suffering to the environments designated by man. So long as a species remains within the limits of the environmental range to which it can adapt, its well-being is assured". These three descriptions thus encompass the notions of "physical and mental health", "feelings", "harmony with the environment" and "adaptation without suffering" which probably cover the broad field we wish to discuss. Of course the term "suffering" has been used to describe welfare so it is necessary to define "suffering" independently.

The Brambell Report (Command Paper 2836, 1965) listed fear, pain, frustration and exhaustion as examples of states of suffering. This method, of drawing up a list of states, which appear to be easily recognisable, has the disadvantage that some might be missed. Social deprivation of various types might lead to something akin to human "loneliness", lack of general stimulation to "boredom" and so on. Moreover, these examples only cover states experienced by human beings. It is possible that animals may experience other, or additional, states of suffering. For example, many animals show behaviour patterns usually called "fixed action patterns" which are stimulated or "released" by certain key stimuli in the environment. In a natural environment, this mechanism usually leads to a chain of co-ordinated behaviour. In an artificial environment, a particular key stimulus may be missing. Often what happens then is that the chain is broken and the animal continues to perform repeatedly some appetitive piece of behaviour without reaching the consummatory phase. An example is the case of some hens which are not stimulated to sit and show normal nesting behaviour by the surroundings of a battery cage and the locomotory nest-searching phase of pre-laying behaviour is very prolonged (Wood-Gush, 1972). In fact, in this particular case, the motivation is so strong that a state of frustration ensues, but one can imagine other cases where an animal might get caught in a closed loop of stimulus-response-stimulus-response without obvious symptoms of frustration being present. Would that animal be suffering? It is very difficult for human beings to judge, since our behaviour sequences depend to a far lesser extent on fixed action patterns and key releasers.

Because of the difficulties of compiling a list of states of suffering, we prefer to use a very loose working definition, namely, "Suffering means a wide range of unpleasant, emotional states".

The problem associated with each of the classes of evidence which may be used as indicators of well-being or of suffering (for of course they are really opposite sides of the same coin) will now be discussed in turn starting with the least debatable and moving to the more controversial. It is not our intention to discuss what all these indicators might be since this is the main function of many following papers. We simply wish to highlight some of the problems that may be encountered.

HEALTH

It is axiomatic that anything which reduces health will also reduce welfare. Disease and injury are major causes of suffering in farm animals. There would therefore appear to be no problem. However, we would like to raise two issues. The first is that it is going to be argued later in this paper that the most important question in deciding about welfare is "What does the animal itself experience subjectively?" Now normally there will be no conflict between these criteria; a human being who is ill usually feels ill. With regard to animals, (and without pre-empting the later discussion on subjective feelings) we can at least say that animals which are ill usually <u>look as though</u> they are feeling ill. However, in the case of human beings we know that there can be disease present especially in the early stages, without any subjective experience of suffering. For example, breast cancer may reach quite an advanced stage without any suffering being experienced. Presumably the same might hold true for animals. Nevertheless, we would argue that although subjective feelings are important, a reduction in health should take precedence as an indicator of suffering.

The second point we would make is that animals may still suffer despite an external appearance of good health. For example, during periods of alarm such as when animals are captured or transported, it may be possible to detect physiological disturbances such as changed hormone levels, which, (as we shall see later), are considered to be indicators of reduced welfare, without any obvious decrement in health (Baldwin and Stephens, 1973; Hails, 1978; Kilgour and de Langen, 1970; Wood-Gush, Duncan and Fraser, 1975). In addition, animals may show behaviour patterns which, (as we shall see later), may be indicative of reduced welfare, and still appear healthy (Duncan and Wood-Gush, 1972; Hediger, 1964; Kiley-Worthington, 1977; Meyer-Holzapfel, 1968). Thus, although ill-health denotes suffering, its absence should not be taken as proof of well-being.

In any discussion on disease and injury the question of pain arises. It is a complex subject not yet fully understood in man let alone in agricultural animals. In a recent comprehensive review of the subject, Bowsher (1981) tentatively concluded that as far as chronic pain is concerned, farm livestock may or may not consciously perceive it, but suffering in connection with it is extremely unlikely. Nevertheless, the physiological mechanisms of pain perception are similar in man and agricultural animals. The behavioural signs to acute pain are also similar in both, namely squealing, struggling, convulsions etc (Command Paper 2641, 1965). This suggests that animals probably suffer from acute pain as we do.

PRODUCTIVITY

A common argument often put forward in the welfare debate is that the high productivity of intensively managed farm animals shows that they cannot possibly be suffering. The fallacies of this argument have been pointed out in the past (Bryant, 1972; Loew, 1972; Ewbank, 1973) but bear repeating again. "Productivity" may mean many different things even in the case of one product, say milk, such as the milk yielded by a cow in a lactation or the milk yielded by a cow per unit of food intake or the economic return from the milk yielded (which might take into account the butterfat and solids content of the milk) per unit of capital, or per unit of labour, or per unit of land. It may be calculated on the basis of an individual animal or of a herd or of a farm. It becomes even more problematical when other products are considered. To take an extreme example, what about the fatty livers from geese? In fact, the use of productivity as a criterion for welfare does have some scientific basis. There is evidence demonstrating the catabolic nature of the stress response with regard to protein (Brown, 1967; Draper, 1967; Baxter and Forsham, 1972). Therefore, if it could be shown that an animal was laying down protein in its tissues, or producing milk or eggs or wool without depleting its bodily protein, to the maximum of its genetic potential, then it would follow that it was not stressed and its welfare would probably be adequate (Duncan, 1981). But what we have just described is a demonstration of a positive nitrogen balance which is a laboratory test and not a simple measure of production or productivity. There are thus many pitfalls in using productivity as an indictor of welfare.

PHYSIOLOGICAL AND BIOCHEMICAL CHANGES

Other papers are dealing with this topic in much more detail so we shall restrict ourselves to considering the problems of relating the physiological and biochemical evidence to other classes of evidence. It has always been hoped that physiological measurements of stress (Selye, 1950) such as arousal of the sympathetic nervous system, release of catecholamines from the adrenal medulla, increased activity of the pituitary and adrenal cortex, and increased levels of plasma gluco-corticoids would give a truly objective measurement of welfare. Many of these measurements can now be made. Some of the original problems are being overcome as scientists refine their techniques. For example, very small permanently indwelling cannulae mean that blood samples can be withdrawn without disturbing the animals (Beuving and Vonder, 1977, 1978), improved micro-analytical techniques mean that blood samples can be minute and the development of radiotelemetry techniques allow certain physiological measurements to be taken with a minimum of interference to the animals (Duncan and Filshie, 1979).

However, there is still the problem of deciding how much of a physiological change an animal can tolerate before we can say it is suffering. The changes that take place are not reflections of on-off switches in particular systems. The sytems are dynamic. Many changes occur as quite normal features of an animal's daily life. Also, it should be remembered that these systems have evolved to help the animal to adapt to changes. They may therefore indicate that the animal is coping with the situation. The biggest problem is therefore how to relate what is measured to whether the animal is suffering.

BEHAVIOUR

The idea that an animal's behaviour might be used to assess its welfare is appealing. If an animal behaved in a particular way when it was suffering, its welfare could be assessed by simple observation without the need for special techniques and without interfering with it.

One possible method of using behaviour as an indicator of welfare is to look for abnormal behaviour. By "abnormal" is meant a persistent, undesirable action, shown by a minority of the population which is not due to any obvious neurological lesion and which is not confined to the situation that originally elicited it (Broadhurst, 1960; Fox, 1968; Meyer-Holzapfel, 1968). Others have said that in addition it should be maladaptive or damaging to the animal (Fraser, 1968).

Once again the problem is to know how much abnormality has to be shown to indicate suffering. There is a very wide range of behaviour patterns which have been called "abnormal". At the extreme end are patterns which involve self-mutilation such as toe-pecking in chicks and since these lead to physical damage and even death, most people would agree that they are symptomatic of suffering. There are others such as mass hysteria which result in injury and death. Again there would be almost universal agreement that these animals were suffering. However, it should be pointed out that mass flight (which is all hysteria might be) could be quite adaptive to escape from a predator in a natural environment. Some behaviour patterns involve the mutilation of others such as cannibalism in pigs and poultry. Although this is a horrifying pattern to observe nevertheless it is of theoretical interest. The injured animals are obviously suffering according to the criterion of a reduction of health. However, they often do not show the behavioural symptoms of pain (avoidance, squealing, struggling) indicating that there may be some mechanism working to spare them the full agony of the situation. It is more controversial whether the animals which are inflicting the damage are also suffering. A hen which is pecking out the entrails of another may be satisfying a need to peck at a worm and a pig which is chewing another's tail may be subjectively chewing a root. It may be the hens which are not pecking worms or entrails and the pigs which are not chewing roots or tails that are suffering!

When less severe forms of abnormality are considered, making judgement on welfare becomes more difficult. However, there should be no disagreement that depraved appetites, usually brought about by dietary deficiencies, indicate suffering. But what of behaviour patterns such as coprophagy and masturbation? Then there is the huge and diverse category of behaviour patterns called "stereotypies". As has been pointed out, where these have been investigated systematically, they appear to have very different causal factors and, by implication, perhaps also very different states that accompany them some of which may involve suffering (Dawkins, 1980).

Another approach using behaviour as an indicator of welfare has been to compare the behaviour of animals on two systems one more intensive or artificial than the other. The assumption is made that the behaviour shown in the less intensive system is in some way more "normal" and that in the more intensive system is indicative of reduced welfare. However, the argument is obviously circular and it should be of no surprise that animals behave differently in different environments. This may simply demonstrate how adaptable they are. These results are only suggestive of reduced welfare. Until it is shown independently that the observed results are indicative of reduced welfare, they will remain inconclusive (Duncan, 1981).

A third method has been to stress animals experimentally by subjecting them to many different types of frustrating or frightening situations and drawing up a catalogue of the behaviour patterns shown. This has been done with domestic fowls and we now know what sort of situations lead to frustration and how the birds respond (Duncan and Wood-Gush, 1971, 1972) and similarly with regard to fear (Jones, 1977a, b, c; Jones, Duncan and Hughes, 1981; Murphy, 1977; Murphy and Wood-Gush, 1978). Of course, it leaves unanswered the question "Are birds which are frustrated or frightened actually suffering?" In fact, there is some evidence from these experiments that severe frustration is aversive (Duncan and Wood-Gush, 1974). Apart from that the results enable more definitive statements to be made about whether or not birds in commercial conditions are frustrated or frightened. But perhaps more importantly they lay the foundation for the next step which is to investigate how birds feel about being frustrated or frightened.

SUBJECTIVE FEELINGS

When we ask whether or not animals are suffering, we really want to know if they are having a particular type of unpleasant mental experience, if they are conscious of what is happening to them, if they are capable of subjective feelings. After all, much of the behaviour that we normally think of as being indicative of reduced welfare could be entirely reflexive with no associated mental experience or awareness of what is going on. It should be remembered that even in the case of human beings, behaviour and subjective experience are not inexorably linked. Bowsher (1981) cites the example of lightly anaesthetized dental patients who may jump, writhe and cry out but subsequently be completely unaware that a tooth has been extracted.

The question of animal awareness has aroused much interest recently and there is universal agreement amongst the reviewers that animals (or at least those as phylogenetically advanced as farm livestock) do not behave like pre-programmed automata but have some sort of awareness (Griffin, 1976; Dawkins, 1980; Wood-Gush, Dawkins and Ewbank, 1981). The problem is, how much? It has been pointed out many times that subjective feelings are not directly accessible to scientific investigation but that we can gain a lot of information about them from indirect evidence. The simplest form of this indirect evidence comes from allowing animals to choose various aspects of their environment. The idea is that animals will express at least some of their feelings in their actions and choose in the best interests of their welfare.

Both Dawkins (1976, 1977, 1978) and Hughes (Hughes, 1975, 1977; Hughes and Black, 1973) have developed these techniques for studying preferences in domestic fowls. There are problems in interpreting the results of these tests (Duncan, 1978b). For example, a choice between two environments only tells us something of their relative properties and not their absolute properties. This can be partly overcome of course by giving a series of choices with a wide range of environments. Also, care must be taken in interpreting results. For example, if a particular set of conditions is consistently and strongly avoided, then it may be assumed that the animal might suffer under these conditions. If, on the other hand, the choice is only a minority one and the animal chooses, say, to spend 10% of its time under one set of conditions then this might be a positive choice and as important for the welfare of the animal as the 90% choice. Another problem is that an animal's short-term preference may not be the same as what it would choose in the long run and may not be in the best interests of its long-term welfare. It is certainly true that animals do not always choose what is best for their physical health. As long as scientists are aware of these limitations, then preference tests will have an important role to play in giving indications of welfare. Their biggest shortcoming at the present is that they have not been used widely enough.

A development of the preference test is to use the fact that animals will learn to perform a response in order either to gain a reward or to avoid a punishment. Theoretically then, we can use operant conditioning procedures to ask animals what they find rewarding and what they find aversive. By making the simple assumption that animals experience subjective feelings of pleasure in the presence of rewards and that they experience distress or suffering in the presence of punishments, it has been suggested that such techniques could provide an objective way of finding out what animals are feeling (Dawkins, 1980).

As yet this method is in its infancy but it shows great promise. The first experiments used simple rewards of some change in the physical environment (e.g. Baldwin and Ingram, 1967; Baldwin and Meese, 1977). However there is no reason why, with some imagination, much more complex situations cannot be investigated. We might be able to ask such questions as "Will an animal work in order to gain access to a particular social group?", "Will it work in order to see or communicate with a group?", "Will it work in order to gain access to a particular environment in which it can perform a certain behaviour pattern?", "Will it work in order to avoid being frustrated in a certain way?" and as a bonus we can also ask in all such cases "How hard will it work?"

CONCLUSIONS

Both well-being and suffering are subjective states which cannot be investigated directly. Because the indicators available are indirect,

we should try to find as many sources of corroborating evidence as
possible. Although disease and injury can be used as indicators of
suffering, their absence is not sufficient to prove well-being. The
problem with physiological, biochemical and behavioural indicators is
one of calibration; how much of a change indicates suffering?
Preference tests have certain limitations and pitfalls but they and
operant conditioning techniques hold out the best hope for the future
for asking animals how they feel.

REFERENCES

Baldwin, B.A. and Ingram, D.L. 1967. Behavioural thermoregulation in
    pigs. Physiol. Behav., 2, 15-21.
Baldwin, B.A. and Meese, G.B. 1977. Sensory reinforcement and
    illumination preference in the domesticated pig. Anim. Behav.,
    25, 497-507.
Baldwin, B.A. and Stephens, D.B. 1973. The effect of conditioned
    behaviour and environmental factors on plasma corticosteroid
    levels in pigs. Physiol. Behav., 10, 267-274.
Baxter, J.D. and Forsham, P.H. 1972. Tissue effects of glucocorticoids.
    Am. J. Med., 53, 573-589.
Beuving, G. and Vonder, G.M.A. 1977. Daily rhythm of corticosterone in
    laying hens and the influence of egg laying. J. Reprod. Fert.,
    51, 169-173.
Beuving, G. and Vonder, G.M.A. 1978. Effect of stressing factors on
    corticosterone levels in the plasma of laying hens. Gen. comp.
    Endocr., 35, 153-159.
Bowsher, D. 1981. Pain sensations and pain reactions. In "Self-Awareness
    in Domesticated Animals" (Ed. D.G.M. Wood-Gush, M. Dawkins and
    R. Ewbank). (U.F.A.W., Potters Bar, England). pp. 22-28.
Broadhurst, P.L. 1960. Abnormal animal behaviour. In "Handbook of
    Abnormal Psychology" (Ed. H.J. Eysenck). (Pitman, London). pp.726-763.
Brown, K.I. 1967. Environmentally imposed stress. In "Environmental
    Control in Poultry Production" (Ed. T.C. Carter). (Oliver and Boyd,
    Edinburgh). pp. 101-103.
Bryant, M.J. 1972. The social environment: Behaviour and stress in
    housed livestock. Vet. Rec., 89, 453-458.
Carpenter, E. 1980. Animals and Ethics. (Watkins, London).
Command Paper 2641. 1965. Report of the Departmental Committee on
    Experiments on Animals. (H.M.S.O., London).
Command Paper 2836. 1965. Report of the Technical Committee to Enquire
    into the Welfare of Animals kept under Intensive Livestock
    Husbandry Systems. (H.M.S.O., London).
Dawkins, M. 1976. Towards an objective method of assessing welfare in
    domestic fowl. Appl. Anim. Ethol., 2, 245-254.
Dawkins, M. 1977. Do hens suffer in battery cages? Environmental
    preferences and welfare. Anim. Behav., 25, 1034-1046.
Dawkins, M. 1978. Welfare and the structure of a battery cage. Size and
    cage floor preferences in domestic hens. Br. vet. J., 134, 469-475.
Dawkins, M.S. 1980. Animal Suffering. (Chapman and Hall, London).

Draper, M.H. 1967. Discussion of part 2 (Physiological and sociological aspects). In "Environmental Control in Poultry Production" (Ed. T.C. Carter). (Oliver and Boyd, Edinburgh). p. 131.

Duncan, I.J.H. 1978a. An overall assessment of poultry welfare. In "First Danish Seminar on Poultry Welfare in Egglaying Cages" (Ed. L.Y. Sorensen). (Nat. Comm. Poult. and Eggs, Copenhagen). pp. 81-88.

Duncan, I.J.H. 1978b. The interpretation of preference tests in animal behaviour. Appl. Anim. Ethol., 4, 197-200.

Duncan, I.J.H. 1981. Animal rights - animal welfare: A scientist's assessment. Poult. Sci., 60, 489-499.

Duncan, I.J.H. and Filshie, J.H. 1979. The use of radio telemetry devices to measure temperature and heart rate in domestic fowl. In "A Handbook on Biotelemetry and Radio Tracking" (Ed. C.J. Amlaner and D.W. Macdonald). (Pergamon Press, Oxford). pp. 579-588.

Duncan, I.J.H. and Wood-Gush, D.G.M. 1971. Frustration and aggression in the domestic fowl. Anim. Behav., 19, 500-504.

Duncan, I.J.H. and Wood-Gush, D.G.M. 1972. Thwarting of feeding behaviour in the domestic fowl. Anim. Behav., 20, 444-451.

Duncan, I.J.H. and Wood-Gush, D.G.M. 1974. The effect of a Rauwolfia tranquillizer on stereotyped movements in frustrated domestic fowl. Appl. Anim. Ethol., 1, 67-76.

Ewbank, R. 1973. The trouble with being a farm animal. New Sci., 60, 172-173.

Fox, M.W. 1968. Abnormal Behaviour in Animals. (Saunders, Philadelphia).

Fraser, A.F. 1968. Behaviour disorders in domestic animals. In "Abnormal Behaviour in Animals" (Ed. M.W. Fox). (Saunders, Philadelphia). pp. 179-187.

Griffin, D.R. 1976. The Question of Animal Awareness. (Rockefeller University Press, New York).

Hails, M.R. 1978. Transport stress in animals: A review. Anim. Reg. Stud., 1, 289-343.

Hediger, H. 1964. Wild Animals in Captivity. (Translated by G. Sircom). (Dover Publications, New York).

Hughes, B.O. 1975. Spatial preference in the domestic hen. Br. vet. J., 131, 560-564.

Hughes, B.O. 1976. Behaviour as an index of welfare. Proc. V Europ. Poult. Conf., Malta. pp. 1005-1018.

Hughes, B.O. 1977. Selection of group size by individual laying hens. Br. Poult. Sci., 18, 9-18.

Hughes, B.O. and Black, A.J. 1973. The preference of domestic hens for different types of battery cage floor. Br. Poult. Sci., 14, 615-619.

Jones, R.B. 1977a. Repeated exposure of the domestic chick to a novel environment: Effects on behavioural responses. Behav. Processes, 2, 163-173.

Jones, R.B. 1977b. Sex and strain differences in the open-field responses of the domestic chick. Appl. Anim. Ethol., 3, 255-271.

Jones, R.B. 1977c. Open-field responses of domestic chicks in the presence or absence of familiar cues. Behav. Processes, 2, 315-323.

Jones, R.B., Duncan, I.J.H. and Hughes, B.O. 1981. The assessment of fear in domestic hens exposed to a looming human stimulus. Behav. Processes, 6, 121-133.

Kiley-Worthington, M. 1977. Behavioural Problems of Farm Animals. (Oriel Press, Stocksfield, England).

Kilgour, R. and de Langen, H. 1970. Stress in sheep resulting from management practices. Proc. N.Z. Soc. Anim. Prod., 30, 65-76.

24

Loew, F.M. 1972. The veterinarian and intensive livestock production, human considerations. Can. vet. J., 13, 229-233.

Meyer-Holzapfel, M. 1968. Abnormal behaviour in zoo animals. In "Abnormal Behaviour in Animals" (Ed. M.W. Fox). (Saunders, Philadelphia). pp. 476-503.

Murphy, L.B. 1977. Responses of domestic fowl to novel food and objects. Appl. Anim. Ethol., 3, 335-349.

Murphy, L.B. and Wood-Gush, D.G.M. 1978. The interpretation of the behaviour of domestic fowl in strange environments. Biol. Behav., 3, 39-61.

Selye, H. 1950. Stress. (Acta Inc., Montreal, Canada).

Wood-Gush, D.G.M. 1972. Strain differences in response to sub-optimal stimuli in the fowl. Anim. Behav., 20, 72-76.

Wood-Gush, D.G.M., Dawkins, M. and Ewbank, R. 1981. Self-Awareness in Domsticated Animals. (U.F.A.W., Potters Bar, England).

Wood-Gush, D.G.M., Duncan, I.J.H. and Fraser, D. 1975. Social stress and welfare problems in agricultural animals. In "The Behaviour of Domestic Animals" 3rd edn. (Ed. E.S.E. Hafez). (Bailliere Tindall, London). pp. 182-200.

DISCUSSION

Chairman: W. Sybesma (Netherlands)

The discussion was focussed upon the two papers of A.Rojahn/FRG (1) and I.J.H. Duncan/U.K. (2).

1. It was referred to the differences existing between the laws of Germany and the other countries of the EC, i.e. the difference between the principle of origin of the animal protection laws. In Germany, the main principle is the ethical aspect towards the animals, whereas in the other countries the laws concentrate on the person who is responsible for the well-being of the animals. That is, animal-oriented aspects versus human-oriented aspects. It was stated that the German law position creates difficulties in finding an unanimously accepted basis for harmonization of the animal protection laws in Europe. Scientists should help in this regard to fill in the gap by providing objective criteria for suffering etc.

Questions were raised whether, in fact, such a difference is essential in protecting the animal. Furthermore, in German animal protection laws responsibility of persons towards the animal has obtained its own place. It was commented that time affects the content of the laws in all European countries. In the past, emphasis was placed on "thou shall", i.e. the responsible human person, nowadays a shift is apparent towards the animals' needs. Thus, the animal-centred approach is put more and more into the laws.

2. In relation to Duncan's paper, a question was asked whether the quality of provoking causes should be taken into account in judging the phenomenon "mass hysteria". This should, indeed, be the case, although it is difficult to give the judgement a label, since the onset of mass hysteria might be a quite normal reaction, such as a sudden flight from a predator. Genetic differences may also be involved.

Further discussions centered around the idea of awareness and its role in preference test experiments. It was suggested that these tests could be very helpful in order to get more objective inside information on the animals' experience of positive or negative feelings or sufferings. Care must be taken, however, that experimental subjects have been raised in "enriched" environments and therefore are better suited to make the appropriate choice.

Suffering as such could also be investigated in a way shown at a seminar on pain, held in the Netherlands in 1981. Too much pain could be measured through behavioural abnormalities in order to obtain some evidence of threshold levels.

3. In the general discussion, the question was raised whether non-material suffering in animals should be taken into account in judging the well-being of animals. The mourning of a dog was given as an example. Comments were given that consciousness and suffering should not be mixed up. For instance, when an animal is unconscious, it may not suffer. However, in such situations one should wonder whether unconsciousness could be proven. Apathy behaviour should not be regarded as only "doing nothing at all". Apathy is a matter of lack of responsiveness and lack of activity. Both these criteria are measurable. In this respect, different environments and changes in the individual play an important role.

In general, it was concluded that also non-material suffering should be taken into account when such suffering could be measured on the basis of changed behaviour.

SESSION II

SIGNIFICANCE OF INDICATORS RELEVANT TO ANIMAL WELFARE

Physiological, biochemical and biophysical
criteria in assessing animal welfare

Chairman: P.V. Tarrant

SIGNIFICANCE OF PHYSIOLOGICAL CRITERIA IN ASSESSING ANIMAL WELFARE

R. Dantzer*, P. Mormède* and J.P. Henry**

*INRA, Neurobiologie des Comportements
Université de Bordeaux II
146 Rue Léo Saignat
33076 Bordeaux Cedex
**University of Southern California
Department of Physiology
School of Medicine
Los Angeles, CA. 90007 USA

ABSTRACT

Many people believe that well-being and suffering are amenable to physiological studies. Evidence is presented to support the idea that when used
in conjunction with behavioural data, physiological variables allow predictions about the ways the animals react to their environment. When measured in
animals acutely exposed to aversive situations, physiological criteria are
reflecting more the intensity of the emotion experienced than its qualitative
aspect. On a long-term basis, however, hormonal activities are differentially
reflecting the different perceptions of the animal. The range of presently
available physiological indicators can be increased by studying neurochemical
changes in the brain or by searching for alterations at the level of effector
organs (e.g. lymphocytes and immune function). In either case, the basic issue
is the relevance of such indicators for animal welfare. Since the use of physiological indicators can only allow to assess the constraints exerted by the
environment on the physical and mental integrity of the organism, there is a
need to define the maximal constraints which are believed to be acceptable.

Physiological alterations are usually taken as evidence of acute stress
for example during transport and handling of animals or as precursors of disease. Under these circumstances, the only problem is to select the appropriate parameter, i.e. the most sensitive one and the easiest to measure. It
is quite different when it comes to assessment of welfare. Welfare refers to
both physical and mental well-being. There is no consensus about where this
state begins and ends. So, the usual strategy to assess welfare is indirect
and consists of looking for evidence of suffering. The difficulty is that
mental suffering is not easier to circumscribe than welfare. One approach to
this problem has been to postulate that suffering arises from excessive arousal of emotions (Brambell, 1965). According to this view, physiological indicators of welfare are confounded with parameters of emotions.

Basic emotional states such as frustration, fear and anger are clearly
accompanied by physiological signs of activation affecting both the pituitary-
adrenal axis and the sympathetic adrenal medullary system. The instantaneous
release of catecholamines and the slower release of glucocorticoids can be

detected by directly measuring the plasma concentrations of hormones or by studying the various changes produced by these hormones at the level of the effector organs (e.g. cardiovascular function, with tachycardia, rise of arterial blood pressure, increased cardiac output ; metabolism, with hyperglycaemia and increase in plasma free fatty acids ; blood cells, with lymphocytosis and eosinopenia, etc...). It is not the aim of this paper to review the different ways of measuring emotions since this knowledge is already available (e.g. Levi, 1975). We will rather address more fundamental questions including (1) the relevance of physiological measurements to specific mental states, (2) the possible ways of increasing the range of physiological indicators, and (3) the key issues in the search for biological indicators of welfare.

1 - Relation of physiological changes to specific motivational states

A positive contribution of physiology to welfare can only be conceived if mental states have distinctive physiological correlates which could therefore be used to indicate the current mood of the animal.

The concept of stress, like most of the theories of emotions, has been dominated by the idea of non specificity. Different stressors (e.g. heat and cold) and different emotions (e.g. joy and anger) lead to similar physiological responses. According to this view, physiological indicators can at best reflect the quantitative dimension of mental states, but not their qualitative characteristics.

Evidence has already been presented to support the idea that non-specificity of the stress response originates from the psychological aspects of environmental stimuli rather than being due to a common biological reaction mechanism. Heat and cold lead to the same physiological response, an increase in the plasma concentrations of glucocorticosteroids, because they are both perceived as potential threat and induce emotional arousal. Indeed, removing the variable of emotional arousal reduces or eliminates the response to most common stressors (Dantzer and Mormède, 1981,1983). This means that it is the psychological rather than the physical representation of the situation in the brain which is of paramount importance to the response.

A related non specific behavioural construct is the concept of arousal. Arousal refers to "a continuum definable operationally from sleep, drowsiness, an inalert state and normal waking through heightened emotional awareness and uneasiness to states of normal emotion, culminating in extreme emotions such as rage, panic and revulsion" (Lader, 1978).

Physiological measures such as heart rate or palmar skin conductance are believed to reflect the arousal state of the organism but not the specific kind of emotion which is experienced. This relation can be obscured by by the fact that the physiological system under investigation is also driven specifically by the metabolic needs of the individual. For example, if an animal is running in a state of panic, the increase in its heart rate will reflect both the arousal associated to panic and the physical effort involved in running. Nevertheless, with adequate precautions and careful choice of physiological measures, valid estimations of arousal can be obtained.

Attempts to induce specific emotions by peripheral injections of adrenaline have led also to the conclusion that the type of emotion experienced by someone depends more on the cognitive cues available in the environment than on the concomitant physiological changes. People who were injected with adrenaline without being informed about the effects produced by the treatment did not feel particularly happy or angry. But when placed with a stooge who acted either in an euphoric way or in an angry manner, they were more prone to show either euphoria or anger than non-treated subjects (Schachter, 1975).

These two lines of evidence explain why a one-to-one relationship between emotional experience and bodily reactions has been dismissed by most authors. However, during the last two decades, research on the mutual relationships between psychological factors and hormonal activities (the discipline known as psychoneuroendocrinology) has shown that this view may hold true only as far as acute emotional states are concerned. On a long term basis, there is a definite relationship between feelings and hormonal profiles, as detailed by Henry and associates (Henry and Stephens, 1977 ; Henry, 1982) : the neurohormonal systems are tuned to respond to the different perceptions of the organism as it, in turn, responds to the demands made on it. Different demands lead to different perceptions which, in turn, lead to different combinations of activation of each of the fundamental neuroendocrine systems.

The sympathetic adrenal medullary system is activated when the power to control access to goal objects such as food, water, shelter, mate and dependents is challenged and leads to repeated attempts to maintain control. This response is associated with continued arousal and increase in heart rate, blood pressure and peripheral/vascular resistance. In contrast, the reverse trend, i.e. deactivation of the sympathetic adrenal medullary system, represents not just absence of effort but relaxation accompanying for example grooming and attachment behaviour (Henry, 1982).

The pituitary-adrenal system is maximaly activated under conditions of uncertainty, when coping attempts are thwarted and potentially disastrous threats are perceived which cannot be escaped or controlled. The subject tends then to act in a passive way, with a predominance of submission and withdrawal. In contrast, security and sense of control are associated with low pituitary-adrenal activity (Henry and Stephens, 1977).

So, on a long-term basis, the major forces which drive endocrine acti-vities are not unpleasant emotions by themselves. The significant factor is the way the subject, animal or human being, perceives himself : loss of con-trol and helplessness are much more important in terms of physiological con-sequences than being angry or fearful. Control implies predictability, i.e. the subject must be able to predict what will happen either independently of what he is doing or as a result of his behaviour ; feedback is also im-portant in the sense that he must be able to determine whether his decision has enabled him to solve the problem or not. Repeated loss of control may lead to helplessness. It is a loss of hope : the subject behaves as if it was no more possible to solve the problem, since he does not see any way out.

Loss of control and helplessness are not the ineluctable outcomes of some set of environmental conditions. They are the result of a transaction between the environment and the way it is perceived by an individual. The same environment may induce loss of control, resignation and ultimately indifference in one individual, loss of control and despair in a second one and repeated attempts to control in a third one. In other words, there is a wide range of individual reactions to the same situation and physiological indicators must give information about the consequences of the situation, but also the propensity for behaving in a certain way in a given situation. This is the difference between what is called in biological psychiatry state markers (e.g. escape during the dexamethasone suppression test in melancholy: Carroll, 1982) and trait markers (e.g. binding of $[^3H]$ imipramine to plate-lets : Langer and Briley, 1981).

All these considerations apply to physiological indicators only as far as they are used in conjunction with behaviour to assess mental experiences. The main characteristic of hormonal activities is that they integrate a wide range of influences from the milieu intérieur and the outside world, so that the same level of functioning may be attained by very different means. For example, catecholamine levels may be up in some pathological states (e.g.

tumor of the chromaffin cells), during the course of an intense physical exercise, during exposure to cold, during a panic attack or because of repeated attempts to maintain control when faced with a very difficult problem. The recourse to sophisticated technology to measure physiological parameters of emotion does not excuse us from the necessity of knowing how the individual is behaving in order to understand what is happening.

2 - The search for new physiological indicators

Physiological indicators are very often classified according to the techniques which are used to measure them (e.g. polygraphy which records any parameter the activity of which generates or can be transduced into an electrical current ; radioisotopic dilution methods which make use of the analogy of behaviour between radioactive labelled compounds and the parent substance, etc.). However, this classification does not tell much about the significance of the parameters under study. A much more useful approach with that respect is to consider physiological events according to their relation to the processing of information which takes place between the perception of environmental stimuli and their effects on bodily processes.

According to this view, hormonal activities are intermediate since they are triggered by nervous events in the brain and they induce a wide range of alterations in effector organs. In first approximation, the specificity of any parameter might be expected to increase the closer it is from the stages involved in central nervous system functioning, since the probability that it will reflect operation of other variables is minimal. A contrario, changes at the level of effector organs will be much less specific since there is a convergent influence of many different physiological systems on these organs. So the search for new indicators can take two opposite directions, by concentrating either on events occurring upstream the endocrine changes or on events occuring downstream. In the first case, the specificity will be higher but the sensitivity lower, while, in the second case, the sensitivity will increase to the prejudice of specificity.

Behavioural neurochemistry is interested in the specific role played by neurochemical brain pathways in the organisation of behaviour. The most intensively studied topics include neurochemical substrates of learning and memory and neurochemical basis of affective disorders and abnormal behaviours. This research has received little application in the field of animal welfare, with the noticeable exception of a few studies on a possible association between oral stereotypies in piglets and abnormalities in dopa-

minergic transmission (Fry et al., 1981). This is a good starting point for
assessing the interest of the neurochemical approach from the point of view
of animal welfare since there is already a substantial amount of clinical
and experimental litterature to support such a working hypothesis. The problem at the present time, however, is that neurochemical measurements are
made post-mortem, so that the study of the dynamics of change in neurotrans-
mission requires too many animals to be compatible with the usual amount
of research funds. Any further development is therefore dependent on the
availability of new analytical tools allowing repeated assessment of neurochemical functioning in vivo. Push-pull cannulae and in vivo voltametry may
permit such measurements, but their feasability has still to be tested.

A non invasive approach to brain functioning in vivo is what has been
called the "neuroendocrine window". Neurochemical pathways modulate the secretion of most pituitary hormones. The functional state of these systems
can be tested by measuring the endocrine response to pharmacological ago-
nists or antagonists of the neurotransmitter under study. For example, catecholamines (noradrenaline and dopamine) modulate the secretion of growth
hormone (GH) and prolactin (PRL). Administration of catecholaminergic ago-
nists increase GH release but decrease PRL release while antagonists have
the opposite action. The amount of change is proportional to the effectiveness of the neuromodulatory control. Although this approach has great potential, neurochemical mechanisms controlling pituitary hormone synthesis and
release are still obscure in farm animals, in contrast with the extensive
amount of work done in laboratory animals and man (Gorewit, 1981).

At the other end of the continuum, changes in the functions of effector
organs may be considered as possible biological indicators. Immunocompetence
is of special interest because of its relation with altered sensitivity to
infectious diseases, which is a common outcome of diverse types of stressors
in animal husbandry (Kelley, 1980). Cell-mediated immune events have been
shown to be altered in a very subtle way according to the time of stress
relative to the induction or expression of the response and the type of
cells involved. This implies that cellular immune events are integrating
more than just the non specific hormonal responses to stress. In addition,
there is evidence that changes in immune functions can occur under conditions of little or no involvement of pituitary-adrenal hormones. For example,
changes in lymphocyte subpopulations and their responsiveness to mitogens
were observed in students submitted to psychological stress (oral examination) in the absence of any significant change in growth hormone, prolactin

and cortisol levels (Dorian et al., 1982).

3 - The true significance of physiological indicators for animal suffering

From the preceding section, it is clear that we are not short of physiological indicators. But the crucial question is what these indicators
are supposed to indicate. As biologists in white coats, we are interested
in understanding the basic mechanisms of interaction between organisms and
their environments and, for that purpose, we try to make use of sensitive
and meaningful parameters. In doing so, we are objective, since we are dealing with material evidence . But when it comes to welfare and suffering, we
are inclined to believe that the question of where to draw the line between
objectivity (based on the above evidence) and subjectivity (the animal right
issues) has to do with ethics and is not a matter of science. At this point,
there is a divorce between the man in his white coat and the everyday man
and, paradoxically, we are inclined to accept just that separation between
psyche and soma ; between the soul and the body that we fight in our research.

The same attitude is leading modern medicine towards a dead end, as
was pointed out in a very sensitive paper by Cassell (1982). The efforts of
physicians tend to aim at curing physical symptoms ; the soul is left to the
church. Concerned only about material evidence and neglecting "immaterial"
matters, medicine has failed to understand the nature of suffering and has
neglected the search for ways to relieve it. Actually in some circumstances
it can even happen, as in a fatal cancer, that the treatment is the source
of more suffering than the disease that it purports to relieve.

In view of this situation, it might seem irrelevant to try to solve
the same problem in the context of animal husbandry. However animal suffering has been given official recognition and has become the object of much
effort to find ways of assessing it and to prevent its occurence. The concern of many biologists for animal welfare is based on the belief that all
ingredients of awareness and consciousness exist in animals (cf. Dawkins,
1980). There has been much more concern, however, about the kind of evidence which can be gathered in favor of the existence of mental states in
animals than about the exact conditions which are responsible for their development and appearance. The types of mental experiences which are eventually accessible to adult laying hens living in battery cages from a very
early age are likely to differ from those of urban or rural pigeons, although the original potentialities of the two species have been pretty much

the same. Whether the well fed and sheltered battery cage bird suffers more than the pigeon scratching for a living in the city is contestable.

Rather than arguing on the kinds of mental experiences which are occurring in different animals housed in different ways, it may be more useful to recognize that animals do indeed experience specific emotions and that, when sufficiently severe, some of these emotions represent suffering. For the good and well-being of animals, it might seem appropriate to prevent them from being exposed to any unpleasant emotional state such as anger, fear or sadness. Maintaining the animal in a permanent state of euphoria would, however, be disastrous on a long-term basis, since it would then have little opportunity to learn how to cope with problems. The implication here is that animals should not be spared all unpleasant emotional states because those are necessary to teach them to adapt to aversive situations when these arise.

The problem here is not only a matter of definition. There are many examples where knowledge and understanding have progressed without any agreement on a definition of the object of study. This is particularly the case with emotions. The language of emotions defines not only inner feelings but also the context in which they arise. So perhaps Mandler is right when he writes that "the best way to study emotions is to ignore them" (Mandler, 1975), i.e. substituting measurable parameters such as behaviour and neuroendocrine changes. While still recognizing that emotions are part of the reaction to specific situations, the task is then to focus on determining what kinds of reactions are occurring and in which way, and what are the critical external and internal factors for their appearance. The major problems we are faced with is not the recognition of emotions or suffering but an agreement on the maximal amount of constraint which can be put on an organism without unacceptable loss of his physical and mental integrity. The decision about what is an unacceptable degree of loss is a responsability that we biologists cannot indefinitely elude.

REFERENCES

Brambell, F.W.K. 1965. Report of technical committee on welfare of animals under intensive husbandry system. (H.M.S.O. London).

Carroll, B.J. The dexamethasone suppression test for melancholia. Brit. J. Psychiat., 140, 292-304.

Cassell, E.J. 1982. The nature of suffering and the goals of medicine. New Engl. J. Med., 306, 640-645.

Dantzer, R. and Mormède, P. 1981. Can physiological criteria be used to assess welfare in pigs? In "The welfare of pigs" (Ed. W. Sybesma). (Martinus Nijhoff, The Hague). pp. 53-73.

Dantzer, R. and Mormède, P. 1983. Stress in farm animals : A need for reevaluation. J. Anim. Sci., accepted for publication.

Dawkins, M.S. 1980. Animal suffering. The Science of Animal Welfare. (Chapman and Hall, London).

Dorian, B., Garfinkel, P., Brown, H., Shore, A., Gladman, D. and Keystone, E. 1982. Abberrations in lymphocyte subpopulations and function during psychological stress. Clin. Exp. Immunol., 50, 132-138.

Fry, J.P., Sharman, D.F. and Stephens, D.B. 1981. Cerebral dopamine, apomorphine and oral activity in the neonatal pig. J. Vet. Pharmacol. Therap., 4, 193-207.

Gorewit, R.C. 1981. Pituitary, thyroid and adrenal responses to clonidine in dairy cattle. J. Endocrinol. Invest., 4, 135-139.

Henry, J.P. 1982. The relation of social to biological processes in disease. Soc. Sci. Med., 16, 369-380.

Henry, J.P. and Stephens, P.M. 1977. Stress, health and the social environment. A sociobiologic approach to medicine. (Springer Verlag, New York).

Kelley, K.W., 1980. Stress and immune function : A bibliographic review. Ann. Rech. Vet., 11, 445-478.

Lader, M., 1978. Current psychophysiological theories of anxiety. In "Psychopharmacology : A generation of progress" (Eds. M.A. Lipton, A. DiMascio and K.F. Killam). (Raven Press, New York). pp. 1375-1380.

Langer, S.Z. and Briley, M. 1981. High-affinity ^{3}H-imipramine binding : a new biological tool for studies in depression. Trends Neuroci., 4, 28-31.

Levi, L. 1975. Emotions, their parameters and measurement. (Raven Press, New York).

Mandler, G. 1975. The search for emotion. In "Emotions, their parameters and measurement". (Ed. L. Levi). (Raven Press, New York). pp. 1-15.

Schachter, S. 1975. Cognition and peripheralist-centralist controversies in motivation and emotion. In "Handbook of Psychobiology". (Eds. S. Gazzaniga and C. Blakemore). (Academic Press, New York). pp. 529-564.

NEUROENDOCRINE STRATEGIES FOR ASSESSING WELFARE :
APPLICATION TO CALF MANAGEMENT SYSTEMS.

Pierre Mormède, Rose-Marie Bluthe, Robert Dantzer

I.N.R.A., Laboratoire de Neurobiologie des Comportements
Université de Bordeaux II
146 rue Léo Saignat - 33076 Bordeaux Cedex

ABSTRACT
    The aim of this paper is to point out the problems associated with the
use of a neuroendocrine strategy to assess welfare in farm animals.We will
first show that methodological problems still obscure our perspective and
that we have not yet done all the experimental work needed to be sure of
the validity of this approach in animal welfare studies. Then,we will present the results of a work aiming at the evaluation of the usefulness of
endocrine parameters to measure the response of young calves to different
rearing systems.

1 - Methodological issues
    The two major neuroendocrine systems which control homeostasis and are
involved in adaptation processes are the hypothalamo-hypophyso-adrenocorti-
cal system (HHCS) and the sympathetic-adrenal medullary system. Both systems are controlled at the hypothalamic level but higher levels of the brain
organization (such as the limbic system or the cortical areas) exerce modu-
latory influences on this control (Dantzer and Mormède, 1983). Two successive phases can be recognized in an animal's response to a stimulus such as
exposure to a new environment : at first it is stereotyped and non specific. Then it progressively vanishes giving place to more subtle changes depending upon the physical and psychological characteristics of the environment. The acute response has been extensively studied in farm animals during
the last decades (Dantzer and Mormède, 1979), but little has been done in
chronic situations. This reflects, in part, methodological problems. The
criteria used to assess the acute stimulation of adrenocortical and sympathetic functions (increased plasma levels of ACTH, glucocorticoid, catecholamines, free fatty acids and glucose, increase, in heart rate, changes
of blood cell counts, etc...) are of no use in the study of chronic influences because their range of variation is within the limits either of spontaneous variations or of alterations induced by the manipulation of the
animal. This problem can to some degree be overcome by special experimental
precautions such 'as chronic venous cannulation to reduce extraneous factors
or by the study of large populations to reduce variability by resorting to

statistics (figure 3). Both solutions obviously have their limitations. It seems more appropriate to look for new criteria in order to assess shifts of endocrine activities under chronic environmental influences.

FIGURE 1.   Enzymes involved in catecholamine synthesis, their main location and regulatory mechanisms.

SYNTHESIS PATHWAY	KEY ENZYMES	LOCATION	REGULATION
Tyrosine ↓ DOPA ↓	Tyrosine hydroxylase	Adrenal medulla Sympathetic ganglia	Nerve impulses
Dopamine ↓ Noradrenaline	Dopamine-beta-hydroxylase	Adrenal medulla , Blood Sympathetic ganglia	Nerve impulses
↓ Adrenaline	Phenylethanolamine-N-methyl transferase	Adrenal medulla	Pituitary-adrenal hormones

Beside their classical effect on hormonal and transmitter release, the foregoing triggering mechanisms have also "trophic" effects on their targets (adrenocortical cells for ACTH, nerve endings or chromaffin cells for nerve impulses). For instance, Kolanowski and coworkers (1975) showed that the adrenocortical response of humans to an ACTH infusion was strongly potentiated by previous stimulation, the response to the fourth of a series of daily ACTH infusions being approximately 3 times larger than the response to the first infusion. In an animal husbandry situation, Friend and coworkers (1979) observed an increased response to ACTH in lactating cows which had had to compete for several days for access to stalls. Such "trophic" influences are also well-known in laboratory animal species with regard to the catecholamine synthesizing enzymes (figure 1). Stress situations such as chronic restraint increase the activity of tyrosine hydroxylase (TH) dopamine β-hydroxylase (DBH) and phenylethanolamine N-methyl transferase (PNMT) in various parts of the vegetative nervous system such as the adrenal gland. This increase proceeds through the synthesis of new enzyme molecules. It has therefore a time lag of 2-3 hours with respect to the stimulation but it is long lasting. TH activity slowly decreases towards prestimulation levels with a half-life of about 3 days (Kvetnansky et al., 1970). Therefore, with successive stimulation, the enzymatic activity will progressively increase until doubled or tripled. Splanchnic nerve activity and corticosteroid secretion are both involved in this phenomenon. Their respective importance depends on the enzyme involved (TH and DBH are principally influenced by nerve activity and PNMT by corticosteroids) and on genetic factors (Ciaranello, 1978 ; Cooper and Stolk, 1979 ; Kvetnansky et al., 1970 ; Stolk and Harris, 1980 ; Wurtman and Axelrod, 1966). Few attempts have been made to evaluate the importance of these measures in farm animals, with the noticeable exception of the work of Stanton and associates in pigs. These authors showed that chronic exposure to a cold environment (Stanton et al., 1972) and early weaning (Stanton and Mueller, 1976) induced a long lasting increase of catecholamine synthesizing enzyme activity in the adrenals and superior cervical ganglion. Further work is needed to assess the value of such criteria as a measure of environmental constraints.

2 - Practical application

The experiments reported here used various measures of endocrine activity to compare two different husbandry systems for young veal calves. The animals under study were commercially bred French Friesian and Norman male and female calves. These animals (two weeks old on the average) were brought from the farm of origin to the fattening unit via a transit center. There, one group was housed in two air-controlled rooms where the animals were on a wooden slatted floor and individually tethered with a chain round their neck, in pens of fourteen. Small wooden bars separated adjacent animals. Another group was housed in an adjacent openshed divided into 8 straw yards each containing 8 calves. Each pen measured 3m x 4m. The front of the shed was protected from the wind by straw bundles and a removable nylon wind screen. Both groups were fed milk substitute in individual buckets twice a day. Two different nutritional patterns were compared during the first two weeks of fattening : a restricted diet normally used to avoid gastrointestinal disorders during the early fattening period and a supplemented

diet. We have shown in a previous experiment that the standard diet failed
to cover the increased metabolic needs induced by transportation (Mormède
et al., 1982). Weight gain during the first two weeks (figure 3) was significantly influenced both by diet (F = 63.21 ; d.f. = 1,116 ; P < 0.001) and
housing condition (F = 16.15 ; d.f. = 1,116 ; P < 0.001). Growth was better
in calves reared in groups. These differences disappeared later.

One week after the beginning of the fattening period, blood cortisol
levels were higher in tied calves (F = 21.91 ; d.f. = 1,107 ; P < 0.001)
together with higher glucose levels (F = 159.43 ; d.f. = 1,99 ; P < 0.001).
The influence of the diet was only marginally significant (Cortisol F = 3.66 ;
P < 0.10 ; glucose F = 4.39 ; P < 0.05). These results clearly show that
housing conditions influence production traits and also those parameters
classically considered to reflect the nutritional status of the animal.

After six weeks of differential housing, sixteen male calves were submitted to an ACTH stimulation test. Animals received either synthetic ACTH
1-24 (Synacthen R), 0.25 mg or saline i.v. immediately after the first
blood sampling. Other blood samples were taken 10, 30, 60 min later. Results
are presented in figure 4. A 3-way analysis of variance with repeated measurements revealed a significant effect of treatment (ACTH versus saline :
$F(1,12) = 124.4$ p < 0.001) housing condition (loose versus tethered :
$F(1,12) = 8.50$ p < 0.05), time ($F(3,36) = 9.50$ p < 0.001) and treatment
x time interaction ($F(3,36) = 7.58$ p < 0.001). Cortisol levels were increased both after ACTH and after saline injection, but with a different
time course and intensity : cortisol levels went up higher after ACTH injection than after saline injection and they came back to initial values at
60 min in animals injected with saline, but remained elevated in animals
injected with ACTH. The difference in basal cortisol levels was no longer
evident at this stage of fattening but tethered calves exhibited a greater
adreno-cortical hormone response than loose calves, whatever the treatment.

The calves were slaughtered on the 105th day of fattening. At that
time left adrenals were taken and hormone levels and enzymatic activities
were measured both in cortical and medullary tissue samples (figure 5).
Higher levels of tyrosine hydroxylase and catecholamines were found in the
medullary tissue but higher levels of corticosterone were present in the
cortical portion. No significant influence of the housing was found in any
parameter. We can only speculate at this time on these negative results.
A possible interpretation is that changes in endocrine activities vanished
with time, due to habituation of the animals to their respective housing

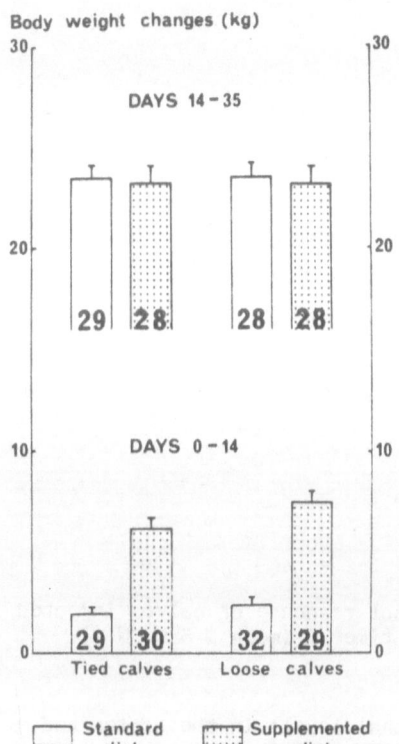

Body weight changes (kg)

DAYS 14-35

DAYS 0-14

Tied calves   Loose calves

Standard diet   Supplemented diet

Fig. 2   Mean body weight changes of the calves during the 5 first weeks after introduction in the fattening unit, according to the diet and housing condition (mean ± SEM).

Fig. 3   Blood cortisol and glucose levels measured one week after the introduction of the calves in the fattening unit, according to the diet and the housing condition (mean ± SEM).

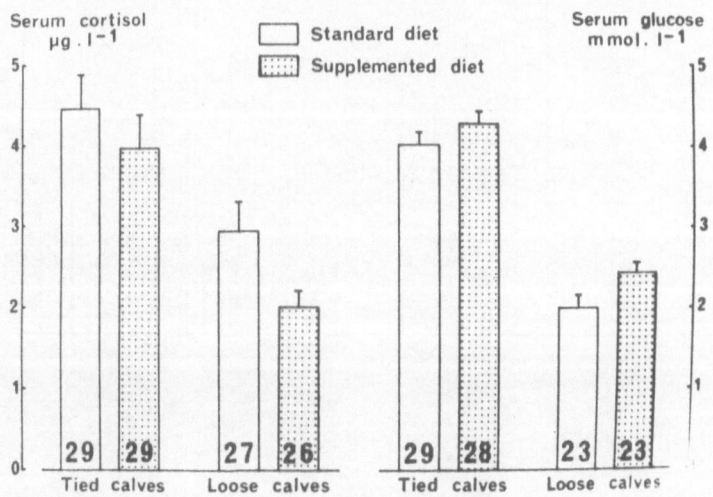

Serum cortisol μg.l⁻¹

Serum glucose mmol.l⁻¹

Standard diet   Supplemented diet

Tied calves   Loose calves   Tied calves   Loose calves

44

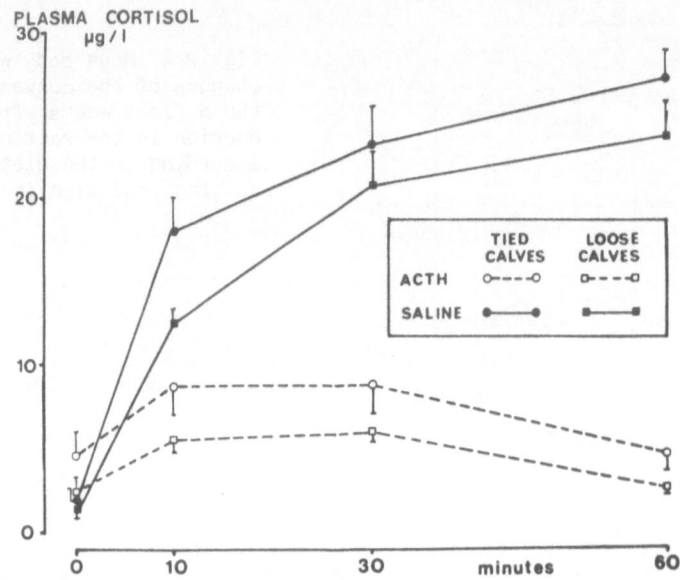

Fig. 4 Kinetics of the plasma cortisol response of calves injected i.v. either 0.25 mg ACTH or saline at time 0 (mean ± SEM, N=4).

Fig. 5 Enzymatic activities and hormonal levels in the cortex and medulla of 17-week old calves (mean ± SEM, N=16).

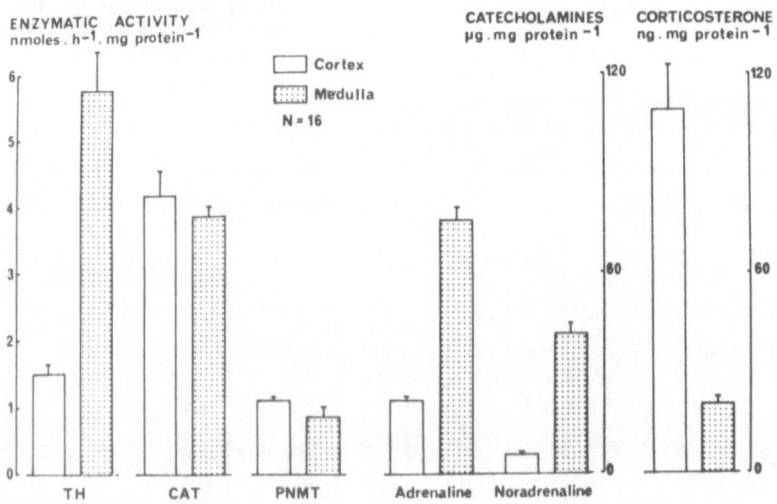

condition, but a definitive answer will be obtained by longitudinal studies with repeated sample testing.

This experiment illustrates an endocrine strategy which could be fruitful in the evaluation of the pressure exerted by the environment on the organism. A combination of different methodologies and experimental conditions help us to understand the various changes. For instance, the higher glucose levels observed during the early fattening period in tied calves is not related to nutritional factors but rather to enhanced catabolism. This condition results in a lower weight gain and is due to a heightened activity of the adrenocortical axis and probably also of the sympathetic system. The second point we want to emphasize here is the importance of longitudinal studies. Adaptation is a dynamic process. Introduction to a new environment elicits in every case a neuroendocrine activation, the novelty being itself a very strong activator of the endocrine system. We hypothesize that the extent and duration of the changes are dependent on the animals' perception of the intensity of the environmental pressure, the normal outcome being an eventual return towards initial activity. Long term shifts would have to be interpreted in terms of specific metabolic needs or psychoneuroendocrine mechanisms. Obviously, the different tools available up to now for these studies have not the same usefulness and sensitivity at each stage of the adaptative process. This is the object of our current research.

Many thanks are due to Jacques Soissons, D.V.M., for the management of the calf studies and to Yvette Langlois for her valuable technical assistance.

REFERENCES

Ciaranello, R.D. 1978. Regulation of phenylethanolamine N-methyltransferase. Biochem. Pharmacol., 27, 1895-1897.

Cooper, D.O. and Stolk, J.M. 1979. Differences between inbred rat strains in the alteration of adrenal catecholamine synthesizing enzyme activities after immobilization stress. Neuroscience, 4, 1163-1172.

Dantzer, R. and Mormède, P. 1979. Le stress en élevage intensif.(Masson, Paris).

Dantzer, R. and Mormède, P. 1983. Stress in farm animals : a need for reevaluation. J. Anim. Sci., accepted for publication.

Friend, T.H., Gwazdauskas, F.C. and Polan, C.E. 1978. Change in adrenal response from free stall competition. J. Dairy Sci., 62, 768-771.

Kolanowski, J., Pizarro, M.A. and Grabbi, J. 1975. Potentiation of adrenocortical response upon intermittent stimulation with corticotropin in normal subjects. J. Clin. Endocrinol. Metab., 41, 453-465.

Kvetnansky, R., Weise, V.K. and Kopin, I.J. 1970. Elevation of adrenal tyrosine hydroxylase and phenylethanolamine-N-Methyl Transferase by repeated immobilization of rats. Endocrinology, 87, 744-749.

Mormède, P., Soissons, J., Bluthe, R.M., Raoult, J., Legarff, G., Levieux, D. and Dantzer, R. 1982. Effect of transportation on blood serum composition, disease incidence and production traits in young calves. Influence of the journey duration. Ann. Rech. Vet., accepted for publication.

Stanton, H.C. and Mueller, R.L. 1976. Sympathoadrenal chemistry and early weaning of swine. Amer. J. Vet. Res., 37, 779-783.

Stanton, H.C., Mueller, R.L. and Bailey, C.L. 1972. Adrenal catecholamine levels and synthesizing activities in newborn swine exposed to cold and 6-hydroxydopamine. Proc. Soc. exp. Biol. Med., 141, 991-995.

Stolk, J.M., Harris, P.Q., 1980. Differentiation of adrenomedullary catecholamine synthesizing enzyme responses to repeated immobilization in hybrid rats. Life Sci., 26, 2099-2104.

Wurtman, R.A. and Axelrod, V. 1966. Control of enzymatic synthesis of adrenaline in the adrenal medulla by adrenal cortical steroids. J. Biol. Chem., 241, 2301-2305.

CORTICOSTEROIDS IN WELFARE RESEARCH OF LAYING HENS

G. Beuving

Spelderholt Institute for Poultry Research
Ministry of Agriculture and Fisheries
7361 DA Beekbergen, The Netherlands

ABSTRACT

The influence of stressful stimuli like heating, handling, withholding food and water, crating, on corticosterone levels in plasma of laying hens had been discussed. A threshold of 5-7 ng/ml plasma seems to exist.
Pre-laying also causes an increase in corticosterone. This increase is the same, when eggs are laid with or without a nest. Comparison of corticosterone increase during laying without a nest the 1st and the 40th egg showed no indication for the development of the disturbed pre-laying behaviour into a stereotypie.

INTRODUCTION

In his formulation of the "General Adaptation Syndrome" Selye (1956) recognized the important role of glucocorticosteroids in coping with stressors. He stated that the action of the adrenal cortex in producing these hormones was a part of the non-specific reactions of the organism to stressful stimuli. Although the concept got some criticism the important role of corticosteroids in coping with stressors is still accepted.

We could ask then if stressful stimuli can be defined in terms of glucocorticoids:

1. Has every increase in corticosteroid levels to do with stress.
2. Means the absence of an increase in corticosteroids also an absence of stress.
3. Does every stressor induce the same corticosteroid response.

Because question 1 and 2 have to do with the definition of stress they can hardly be answered. We will meet this problem, however, in the interpretation of some of our results. In our Institute most of the research on this topic has to do with question 3: do stressors induce corticosterone responses in laying hens?

TECHNIQUE

All blood samples were taken by a cannula in the wing artery, to prevent any influence of handling. The birds with the cannula could move freely in their cage, also during blood sampling.

When many blood samples were taken from one hen, extra blood was supplied. This blood was obtained from cannulated "donor" hens and was injected directly via the cannula.

Care must be taken that the blood flow is not too fast during blood sampling, because a haemorrhage induces an increase in ACTH levels that is not suppressible by corticosteroid feed back (Gann et al., 1978).

Corticosterone was determined by the method of competitive protein binding in the experiments of application of stressors (handling, heat, crating, etc.). In all other experiments a radioimmunoassay was used. There is a strong correlation between these assays (Beuving and Vonder, 1981).

BASELINE LEVELS

Baseline levels of corticosterone in birds were low as compared with mammalian levels. In laying hens the mean daily fluctuations (the daily rhythm)were in the range of 0-2 ng/ml plasma. The individual variation was determined by taking blood samples with short intervals (2 minutes). Sometimes a regular pattern is found with a periodicity of about 12 minutes, while sometimes a more irregular pattern was shown. This seems plausible when we realize that the amount of corticotropin releasing factor - not yet isolated - is the net result of a summation process and depends on a number of inputs like rhythmic factors, eating and drinking activities, as well as emotional disturbances, temperature changes, tissue damage, bacterial infections, etc.

Because of the great variation of corticosterone in baseline levels and also during stress and the great variation between individuals, the number of experimental birds cannot be low. We tried to have at least 10 animals in each experimental group.

APPLICATION OF STRESSORS

When we applied stimuli, that were thought to be stressful, we found increases in nearly all cases (Beuving and Vonder, 1978). These stimuli were more physical (like temperature increase, withholding food and water) or more psychological (handling, repeated handling, crating, removing a laying nest).

An important question is the level of corticosterone reached after application of each stimulus: is this level dependent of the nature of that stimulus? There are indications in literature that the increase in corticosterone would be independent of the stimulus character. Natelson et al.(1981) found an increase in corticosterone levels in rats, when footshock was given; but they found no relationship between the strength of current and the level of plasma corticosterone. However, Dallman and Jones (1973) did found - also in rats - a linear relationship between corticosterone levels and voltage of the footshock. But de Souza and van Loon (1982) stated that a discrete stress is associated with a specific limited peak response. Application of the same stressor after a short interval did not further increase the level reached by the first application of this stressor.

In our experiments we found some differences in corticosterone response between the applied stressors. However, most stimuli did increase hormone levels to values of about 5-7 ng. Handling during 7 minutes causes a very strong increase (13 ng/ml) but this might be mainly brought about by haemorrhage. Crating for 7 hours or immobilization by hand for 1 hour causes a fast increase to a level of about 5-7 ng/ml and next a gradual increase to values of about 10 ng/ml. More physical stimuli like heating and deprivation of food and water induced the same levels.

One could assume that the limited response after application of most stressors was caused by the limited capacity of the adrenal to synthetize corticosterone. However, injection of porcine ACTH produced high levels of corticosterone (30 ng/ml) suggesting a high adrenal potentiality.

These results show that several levels can be reached under different conditions; but it seems as if a threshold of 5-7 ng/ml exists.

EGG LAYING AND CORTICOSTEROIDS

The effect of egg laying has been studied more intensively. When laying hens were housed individually with a laying nest, they perform a rather normal pre-laying behaviour, spending most of the time sitting on the nest. Nevertheless a distinct rise in corticosterone, peaking at the time of oviposition, could be shown. When laying hens were caged individually without a laying nest, the pre-laying behaviour is disturbed (restlessness, pacing and escape behaviour). However, the same increase of corticosterone levels was found (Beuving and Vonder, 1981). By separation the actual egg laying from the pre-laying behaviour it could be shown that the increase of corticosterone was coupled to pre-laying behaviour.

So the question arises what might be the cause of the increase in corticosterone levels during this behaviour.

One explanation might be that the disturbed pre-laying behaviour
ritualised into stereotypic movements. The increase of corticosterone
should then gradually diminish in the process of ritualization. To check
this hypothesis a comparison was made between the increase of corticosterone levels during laying - without a laying nest -
1. the first egg
2. at least the 40th egg.

The results do not support this hypothesis (Fig.1). The only difference found was the much longer lasting increase (about 1 hour) of corticosterone levels during pre-laying behaviour of the first egg.

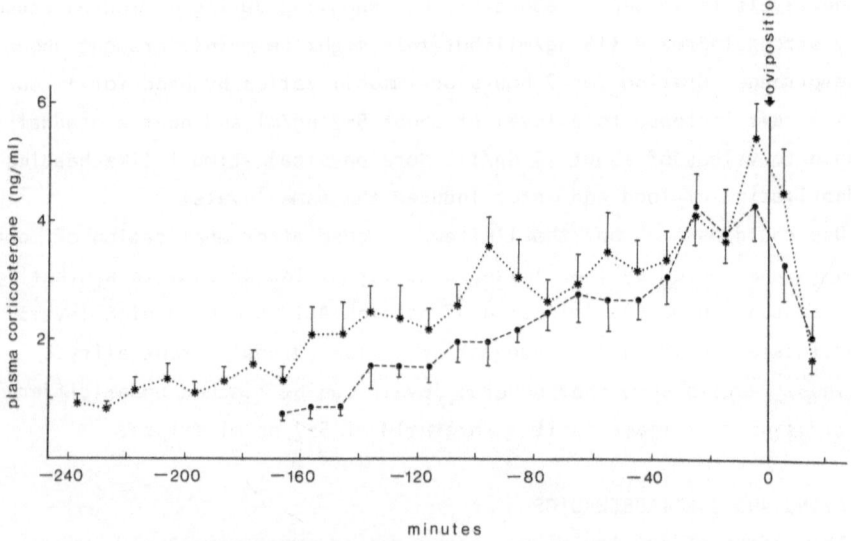

Fig. 1   Concentrations of corticosterone (+ S.E.M.) before
         oviposition during the 1st egg and the 40th egg.
         Blood samples were taken each 5 minutes.
         .... 1st egg N=10
         ---------------- 40th egg N=10

One might wonder if there is any influence at all of the presence of
a laying nest on the increase in corticosterone levels. Therefore we performed another experiment. Laying hens, individually housed in cages, were
given access to a laying nest from the onset of lay. After at least 40
eggs were laid the increase in corticosterone levels during egg laying was

measured the day before and the day after the nest was removed. Under these circumstances an extra increase in corticosterone levels could be shown (Fig.2).

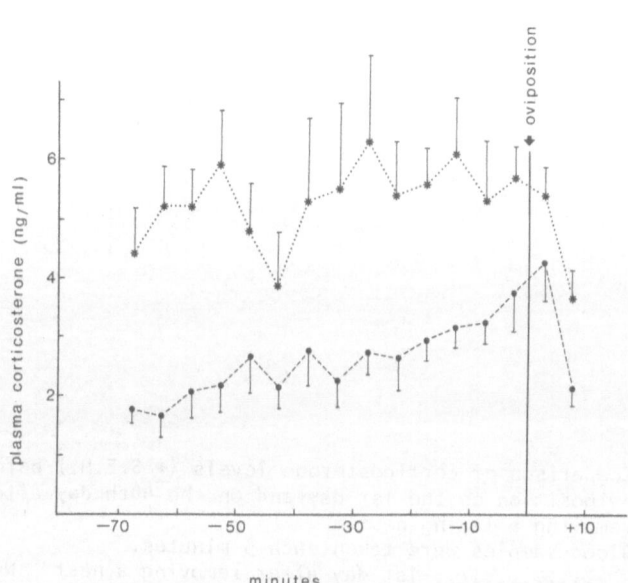

Fig. 2 Effect of removing a nest on corticosterone levels (± S.E.M.) in plasma before oviposition. Blood samples were taken each 5 minutes.

--------------- day before removing a nest N=8
.... day after removing a nest N=8

This extra increase, however, disappeared after another 40 eggs were laid without a laying nest (Fig.3).

The extra increase of corticosterone can be caused by the absence of a nest (novelty) or by the disturbance of pre-laying behaviour. The first reason (novelty) looks the most probable cause of stress, since we found no effect of disturbed pre-laying behaviour on corticosterone levels.

It is difficult to find a fitting explanation for the increase of corticosterone during egg laying. On the one side we do not know the function of corticosteroids in regulation of pre-laying behaviour.

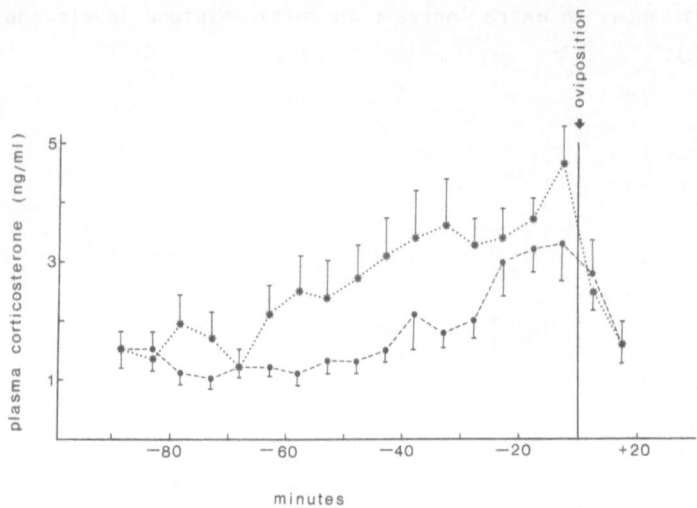

Fig. 3 Comparison of corticosterone levels (+ S.E.M.) before oviposition on the 1st day and on the 40th day after removing a laying nest.
Blood samples were taken each 5 minutes.
.... 1st day after removing a nest   N=9
-------------- 40th day after removing a nest   N=9

On the other side we do not see a clear relation with other hormones during this stage. Progesterone and estrogen are involved in egg laying behaviour (Wood-Gush and Gilbert, 1975), but no short interval concentrations of these hormones in plasma are determined. So no indications are available for a hormonal interrelationship.

CONCLUDING REMARKS

Although corticosterone levels in laying hens showed a response to nearly all applied stimuli, we believe that corticosterone has a limited possibility as a parameter of stress as far as we want to relate the level of corticosterone to the level of stress. Gradation appears to take place in a rather small region of 2-6 ng, although higher concentrations are possible under certain conditions.

To estimate the physiological response after application of stimuli also other parameters (like catecholamines and heart rate frequency) can be included for a better understanding of this response.

Another field of research in future will be on the function of corticosteroids in behavioural responses to see if correlations exist between resting hormone levels or increases of hormone levels and the capacity of the animals to cope with the new situation.

REFERENCES

Beuving, G. and Vonder, G.M.A. 1978. Effect of stressing factors on corticosterone levels in the plasma of laying hens. Gen.Comp.Endocrinol. 35, 153-159.
Beuving, G. and Vonder, G.M.A. 1981. The influence of ovulation and oviposition on corticosterone levels in the plasma of laying hens. Gen. Comp.Endocrinol. 44, 382-388.
Dallman, M.F. and Jones, M.T. 1973. Corticosteroid feed back control of ACTH secretion: effect of stress-induced corticosterone secretion on subsequent stress responses in the rat. Endocrinology, 92, 1367-1375.
Gann, D.S., Ward, D.G. and Carlson, D.E. 1978. Neural control of ACTH: a homeostatic reflex. Recent Progr.Horm.Res. 34, 357-400.
Natelson, B.H., Tapp, W.N., Adamus, J.E., Mittler, J.C. and Levin, B.E. 1981. Humoral indices of stress in rats. Physiol.Behav., 26, 1049-1054.
Selyle, H. 1956. The stress of life. McGraw-Hill, New York.
Souza, E.B.de, and van Loon, G.R. 1982. Stress-induced inhibition of the plasma corticosterone response to a subsequent stress in rats: a noradrenocorticotropin-mediated mechanism. Endocrinology 110, 23-33.
Wood-Gush, D.G.M. and Gilbert, A.B. 1975. The physiological basis of a behaviour pattern in the domestic hen. In "Avian Physiology" (Ed. M. Peaker). (Academic Press, New York, San Francisco, London) pp 261-276.

THE SLEEP-WAKING PATTERN AND BEHAVIOR OF PIGS
KEPT IN DIFFERENT HUSBANDRY SYSTEMS

J. Ladewig and F. Ellendorff

Institut für Tierzucht und Tierverhalten (FAL) Mariensee,
3057 Neustadt 1, Federal Republic of Germany

ABSTRACT

The use of an analysis of the sleep-waking pattern and related behaviors as an indication of adaptation to different housing conditions is presently being investigated in German Landrace pigs. Sleep and wakefulness is determined by means of electroencephalography, and simultaneous behavioral observations are conducted by means of video equipment. The pigs are kept, first, in an open pen with straw, later in a farrowing crate with slatted floor. Preliminary results indicate that, although rooting behavior was greatly reduced after the change in housing system, with a corresponding increase in the time spent lying down inactively, no uniform change occurred in sleep and wakefulness. The possibility that earlier experience with slatted floor types without straw is responsible for the lack of change in sleeping behavior will be investigated in future experiments.

INTRODUCTION

Assessment of adaptation to different housing conditions in farm animals is a difficult task that has not, yet, been solved with satisfaction. It is wellknown, however, that environmental changes are able to affect the sleep-waking pattern of humans (Passouant, 1980), as well as animals (Mirmiran and van den Dungen, 1980; Ruckebusch, 1975). Thus analysis of sleep and wakefulness may be useful to detect gross disturbances due to exposure to novel housing, or to indicate changes occurring during adaptation to a novel husbandry system.

Along these lines, we are presently involved in experiments in which the sleep-waking pattern is analyzed in domestic pigs whoes environment is altered from soft bedding and with the possibility of free movement to slatted floors and restriction of movement. To date, results of the sleep-waking pattern and related behavior from three animals are available.

METHODS

Three castrated male German Landrace pigs weighing 70 -
80 kg were used for the study. For the first half of the
study period, each subject was kept individually in an open
pen with straw. The pen was located in an environmentally
controlled room. Apart from temperature, humidity, and light-
darkness control, the room was noise and vibration shielded,
and the air flow was controlled. Water and feed was provided
ad libitum, and a miniature pig was kept in a separate cage
in the same room to avoid effects of isolation.

For the second half of the study period, the subjects
were kept in a farrowing crate with slatted floor. The crate
was built inside the open pen, after removal of the straw, to
avoid other changes in the environment.

Electroencephalographic and behavioral recordings

An analysis of the sleep-waking pattern was conducted on
the basis of an electroencephalogram (EEG). Silverball electrodes were implanted surgically on the dura of the frontal
cortex. Other electrodes were implanted in the eye and neck
muscles for the recording of an electrooculogram (EOG) and
an electromyologram (EMG). All electrodes were fastened to a
connector which was fixed to the os frontale of the animal.
Immediately after the operation each animal was placed in the
test room and connected on-line to a polygraph located outside
the test room.

Based on the polygraph recordings of the EEG,EOG and EMG,
four different sleep-waking stages could be distinguished.
Two different phases of wakefulness, the wake stage (W) and
the relaxed stage (Re), and two different phases of sleep,
the slow wave sleep stage (SWS) and the paradoxical sleep
stage (PS), were recorded. In the W-stage, the EEG showed
fast fluctuations with small amplitude. The EOG showed volun-
tary eye movements, and the EMG showed tonus in the neck
muscles. In the Re-stage all recordings were similar with
the exception that, in the EEG, brief periods of slower
fluctuations with greater amplitude occurred.

In the SWS-stage the recordings in the EEG were slower but with larger amplitude. The EOG showed absence of eye movements, and the EMG showed little or no tonus in the neck muscles. In the PS-stage the EEG was similar to that of the W-stage. The EOG showed occurrence of regular eye movements (socalled Rapid Eye Movements or REM), and the EMG showed complete lack of tonus in the neck muscles.

Apart from the four sleep and wake stages, four different body positions were registered: Lying on the side, lying on the sternum, sitting, and standing or walking. Furthermore, four different categories of behavior were recorded: Inactivity, rooting behavior, eating, and chewing of objects other than straw.

For the behavioral observation of the subjects a videocamera was located in the test room. In order to synchronize the behavioral observation with the sleep-waking pattern, a second camera was focussed on the polygraph recordings.The pictures from the two cameras were mixed and recorded with time lapse on a videotape for later evaluation.

After an adaptation period of at least 14 days, during which time each subject was kept in the open pen with straw, and equipped with the cable and with all environmental conditions controlled, a series of five tests was given within ten days.Each test consisted of a 24 h observation period.

After the initial five tests, the housing system was changed into the farrowing crate with slatted floor, and another series of five tests was conducted within ten days in a similar manner.

Analysis of results

For each 24 h observation period, the frequency and duration of each sleep-wake stage, body position and behavioral category was registered, and a mean duration calculated. From the first five tests, a mean ($\pm$ SD) was calculated for each category and compared with the mean from the last five tests. Due to the low number of experimental subjects so far, we have not yet performed detailed statistical evaluation of the data.

RESULTS

Body position

As expected, all subjects spent more time lying down in the farrowing crate than in the open pen. For one subject, the increased time of lying was spent primarily lying on the sternum rather than on the side, whereas for another subject the increased time of lying was spent primarily lying on the side. For the remaining subject, the increase in lying duration was equally distributed between lying on the side and lying on the sternum.

Corresponding to the increase in lying time, all subjects tended to spend less time sitting and standing in the cage.

Behavior

The change in behavior induced by the environmental change corresponded to these changes in body position in so far as all subjects spent much less time rooting in the cage with slatted floor than in the pen with straw. Also the frequency of periods with rooting behavior was reduced in all subjects. Consequently, periods of inactivity lasted much longer in the cage than in the pen but occurred less frequently. On the other hand, the frequency and duration of periods of eating or chewing did not differ in the two types of systems.

Sleep-waking pattern

The total time that the subjects were awake or asleep in the two housing systems varied only slightly. One subject spent more time sleeping in the cage than in the pen, but no difference was found in the two other subjects. The three subjects slept, in average, 10 h and 22 min per 24 h period ($\sim$ 43.4 per cent). Of the total sleep time, 70.5 per cent was spent in the SWS-stage and 29.5 per cent in the PS-stage. The PS-stage constituted 12.7 per cent of the whole 24 h period.

During wakefulness, all three subjects spent considerably less time in the W-stage and more time in the Re-stage in the farrowing crate as compared to the open pen.

When the duration of each sleep-waking stage or behavioral category of each individual test was plotted against
test number, no characteristic change was found in two of the
subjects other than what has already been described. For the
third subject,however, a characteristic change, particularly
in the sleep-waking pattern, was seen during the first test
after the environmental change. The duration of total sleep
time was considerably reduced on this test, due to a marked
reduction in both SWS and PS.

On the following tests, however, the total sleep time was
increased, and reached a level higher than that found prior
to the environmental change.

DISCUSSION

Comparison of the behavior and body position in the two
management systems indicates that the subjects appeared to
have less to do in the farrowing crate with slatted floor than
in the open pen with straw, and therefore spent more time
lying down inactively. The resulting decrease in rooting behavior, however, did not seem to affect the sleep-waking
pattern in any uniform way, although particularly rooting behavior under normal circumstances is a highly characteristic
and frequently conducted behavior in pigs.

Studies on the sleep-waking rhythm of production animals
under various conditions, conducted by Ruckebusch (1975), indicate that particularly three aspects of the sleep-waking
cycle may be considered indicators of environmentally induced
disturbance: First, changes in total sleep time, either SWS
alone or SWS and PS. Secondly,no change in total sleep time,
but a reduction of the mean duration of SWS episodes, i.e.
fragmentation of the SWS profile. And thirdly, no variation
in total sleep time or fragmentation, but a delayed onset of
sleep.

In our study, only one subject showed signs of disturbance of the sleep-waking rhythm. On the first test conducted
directly after the change in environment, a marked reduction
in both SWS-time and PS-time, and a corresponding increase in

Re-time indicated lack of adaptation to the farrowing crate with slatted floor. Already in the following 24 h period, however, the sleep pattern corresponded to that of the other subjects, indicating full adaptation. As for the other subjects, no indication of disturbed sleep was found, neither fragmentation of the SWS profile nor delayed onset of sleep.

All three subjects had been raised on slatted floor prior to the study period. It is possible that earlier adaptation to this type of housing system is the reason why the change in environment during the study period had no effect on the sleep behavior. Further research will analyze this possibility in the future.

REFERENCES

Mirmiran, M.and van den Dungen, H. 1981. Influence of a complex ('enriched') environment on the sleep-waking patterns of developing rats. In "Sleep 1980" (Ed. W.P. Koella). (Karger,Basel). pp. 351-353.
Passouant, P. 1980. Pathology of sleep. In "Biology of sleep. An interdisciplinary survey". (Ed. M.Monnier). Experientia, 36, 1-27.
Ruckebusch, Y. 1975. The hypnogram as an index of adaptation of farm animals to changes in their environment. Appl. Anim. Etho., 2, 3-18.

NEUROCHEMISTRY OF ABNORMAL BEHAVIOUR IN FARM ANIMALS

D.F. Sharman

Agricultural Research Council, Institute of Animal
Physiology, Babraham, Cambridge CB2 4AT, U.K.

ABSTRACT

Pharmacological evidence indicates that dopaminergic
neuronal systems in the brain might be involved in abnormal
oral behaviour in farm animals.
Early-weaning of piglets appears to result in a reduction
in the turnover of dopamine in the brain.

INTRODUCTION

There is pharmacological evidence to suggest that
dopaminergic neurons in the brain are involved in certain
forms of abnormal oral behaviour in farm animals. Between
1873 and 1875, Johann Feser, the first Professor of
Veterinary Pharmacology at the Veterinary School in München,
investigated the actions of apomorphine in farm animals.
He concluded (Feser, 1875) that the behavioural effects of
apomorphine closely resembled the symptoms of "licking
sickness" (Lecksucht or pica) and that the part of the brain
that was affected by apomorphine was disturbed   when this
disorder occurred. Feser suggested that apomorphine be used
to treat "licking sickness". Sharman and Stephens (1974)
also observed the behavioural similarity between the effects
of apomorphine in cattle and sheep and abnormal oral behaviour
that had been reported to occur in these species   (Stephens
and Baldwin, 1971; Stephens, 1974). Apomorphine is thought
to induce stereotyped gnawing behaviour in laboratory rodents
by stimulating central dopamine receptors (Ernst, 1967).

In 1962, Bollwahn reported paradoxical behavioural
effects of the tranquillizing drugs perphanazine and
chlorprothixene in pigs. These included compulsive chewing
and biting at the food trough, or the floor or wall of the
pen. We have tested several neuroleptic drugs in pigs and
have confirmed these observations. We have also frequently
observed "snout-rubbing" following treatment of pigs with
neuroleptic drugs. Intense, repetitive "snout-rubbing" is a
characteristic response of pigs to apomorphine (Feser, 1874).

Of the tranquillizing drugs we have examined, metoclopramide, a drug employed in human medicine as an anti-emetic, induced "snout-rubbing" in approximately 33% of the pigs tested. One of the actions of neuroleptic or major tranquillizing drugs is the blockade of dopamine receptors in the brain. They also induce an increased turnover of dopamine in the brain which is reflected by an increase in the cerebral concentrations of the acidic metabolites 3,4-dihydroxy-phenylacetic acid (DOPAC) and 4-hydroxy-3-methoxy-phenyl-acetic acid (homovanillic acid; HVA).

It would thus appear that abnormal oral behaviour resembling the vice of "snout-rubbing" (Allison, 1976) can be induced by drugs which stimulate and by drugs which block, dopamine receptors in the brain. This report reviews our studies on the cerebral metabolism of dopamine in early-weaned piglets showing stereotyped "snout-rubbing".

The metabolism of dopamine in the brains of piglets showing stereotyped "snout-rubbing"

In a study of the effects of reduced oral stimulation associated with feeding in piglets removed from the sow at one day old, it was found that piglets, fed through an indwelling gastric tube showed more stereotyped "snout-rubbing" than litter mates fed by means of a bottle fitted with a rubber teat and allowed to suck at the teat until sated. The piglets fed intragastrically showed a lower concentration of HVA in the putamen and nucleus accumbens, indicating a reduced turnover of dopamine in these brain regions (Fry et al. 1981). Further investigations (Sharman et al., 1982) into the cerebral metabolism of dopamine and stereotyped snout-rubbing behaviour indicated that reduction in the metabolism of dopamine in the brains of early-weaned piglets was associated with separation from the sow. In all of these experiments the early-weaned piglets were kept singly after removal from the sow. The effect of keeping piglets under such conditions has not yet been investigated but Hutchins et al., (1975) showed that in male mice, kept singly for six weeks or more, there was an altered metabolism of dopamine only when such animals were exposed to a fresh environment.

The effects of neuroleptic drugs on behaviour and cerebral dopaminergic neuronal mechanisms in pigs

The short-acting drug, metoclopramide, has been employed to study the abnormal behaviour induced by neuroleptic drugs in the pig. Stereotyped snout-rubbing occurs about 15 min after the drug administration in contrast to some other neuroleptic drugs which can cause such behavioural abnormalities to appear 24h after the administration of the drug. During the first 15 min after receiving metoclopramide the pigs remain quiet and can be easily handled. The onset of the stereotyped snout-rubbing is sudden and the behaviour resembles that induced by apomorphine.

Metoclopramide will prevent the onset of the behavioural effects of apomorphine in pigs and after metoclopramide, the concentrations of DOPAC and HVA, in the caudate nucleus of the pig, are increased. These findings are indicative for the blockade of central dopamine receptors.

Studies on the characteristics of binding sites for neuroleptic drugs on membranes derived from pig caudate nucleus did not reveal any major differences from the characteristics of such binding sites in the rat caudate nucleus (Sharman and Banns, 1980). In the rat, abnormal behaviour following short-term treatment with neuroleptic drugs is rarely seen. The guinea-pig also shows "snout-rubbing" behaviour in response to metoclopramide and in this species it has been found that blockade of central dopamine receptors is necessary for the response to occur (Rodriguez del Camino and Sharman 1981).

From these findings it would seem that the conclusion of Feser (1875) was correct. In piglets showing stereotyped snout-rubbing there is an altered metabolism of dopamine in parts of the brain that receive a dopaminergic neuronal input. Apomorphine is thought to act at dopamine receptors in these brain regions. However, it is not yet clear why apomorphine, which is thought to stimulate dopamine receptors should induce in the pig a behavioural response similar to that caused by drugs thought to block dopamine receptors.

REFERENCES

Allison, C.J. 1976. Snout-rubbing as a vice in weaned pigs. Vet. Rec., 48, 254-255.

Bollwahn, W. 1962. Beobachtungen bei der Anwendung von Chlorprothixen und Perphenazin beim Schwein. Dtsche. Tierärtzlche. Wschr., 69, 219-224.

Ernst, A.M. 1967. Mode of action of apomorphine and dexamphetamine on gnawing compulsion in rats. Psychopharmacologia (Berl.), 10, 316-323.

Feser, Prof. 1874. Die in neuester Zeit in Anwendung gekommenen Arzneimittel 1. Apomorphinum Hydrochloratum. c. Wirkungen desselben beim Schweine. Z.prakt. Veterinairwissenschaft, p. 310-317.

Feser, Prof. 1875. Apomorphinum Hydrochloratium, ein Heilmittel gegen die sog. Lecksucht der Rinder, Schafe und Schweine. Z.prakt. Veterinairwissenschaft, p.111-113.

Fry, J.P., Sharman, D.F. and Stephens, D.B. 1981. Cerebral dopamine, apomorphine and oral activity in the neonatal pig. J. vet. Pharmacol. Ther., 4, 193-207.

Hutchins, D.A., Pearson, J.D.M. and Sharman, D.F. 1975 Striatal metabolism of dopamine in mice made aggresive by isolation. J. Neurochem., 24, 151-1154.

Rodriguez del Camino, I.L. and Sharman, D.F. 1981 Apomorphine antagonizes the stereotyped behaviour produced by metoclopramide in the guinea-pig. Br. J. Pharmacol., 74, 767-768P.

Sharman, D.F. and Banns, H. 1980. Problems associated with neuroleptic drugs in pigs. Devels Neurosci., 8, 495-498.

Sharman, D.F., Mann, S.P., Fry, J.P., Banns, H. and Stephens, D.B. 1982. Cerebral dopamine metabolism and stereotyped behaviour in early-weaned piglets. Neuroscience, 7, 1937-1944.

Sharman, D.F. and Stephens, D.B. 1974. The effect of apomorphine on the behaviour of farm animals. J. Physiol. (Lond.), 242, 25-27P.

Stephens, D.B. 1974. Studies on the effect of social environment on the behaviour and growth rates of artificially reared British Friesian male calves. Anim.Prod., 18, 22-34.

Stephens, D.B. and Baldwin, B.A. 1971. Observations on the behaviour of groups of artificially reared lambs. Res. vet. Sci., 12, 219-224.

DISCUSSION

Chairman: P.V. Tarrant/Ireland

This session dealt with the physiological signals which can tell us when something is wrong with the well-being of the farm animals, and which help us to quantify the problem. The first requirement is to identify the appropriate signals to look for and the second requirement is to interpret the values obtained in terms of animal welfare. If we can achieve that, we are in a position to identify and rank environmental stressors, in terms of welfare.

Progress along this direction can only be achieved with the support of behavioural studies. This is because of the difficulty of distinguishing between acceptable and unacceptable forms of stress (or arousal) both of which may generate some of the same physiological and biochemical signals. Thus, by a logical progression we arrive at the concept of integrated systems of indicators, which will be the subject of Session 3.

The question was raised as to the use and validity of physiological data for the identification of suffering. It was considered that physiological data per se may not be indicative of suffering or well-being, but simply quantifies how an animal is reacting to the environment. Physiological measurements by themselves cannot give a full understanding but must be combined with behavioural data to be meaningful in relation to animal suffering.

It was suggested that some of the difficulties of equating physiological signals might be overcome by the use of operant conditioning techniques, in which the animal could alter its physiological state.

The possibility of pharmacological intervention to give relief from suffering was prompted by D.F. Sharman's (U.K.) paper. Perhaps therapy could extend to other disturbed behaviours in addition to abnormal oral behaviour? Problems raised by the therapeutic use of drugs would include efficacy, expense, whether repeated treatments would be necessary to get an adequate response, and whether other abnormal behaviours could respond to therapy. However, problems of drug residues would almost certainly rule out such an approach in animals produced for food.

Improved techniques for in vivo monitoring of changes in brain chemistry are urgently needed. Present methods which rely on post mortem assay, for example of endorphin levels, are very limited in their application to the study of animal welfare. Neurochemistry offers many possibilities for future advances, particularly with respect to neurotransmitters, endorphins and other neuropeptides. This field of basic research could be very important for welfare studies.

Recently, progress has been made in techniques for sampling neurotransmitters in vivo; "push-pull cannulation" allows sampling of the fluid surrounding the neurons. Voltametry techniques may, in future, allow measurement of the catecholamine neurotransmitters during behavioural studies. Finally, tapping cerebrospinal fluid from the ventricular system of the brain is another useful technique for in vivo studies of brain chemistry and could find application in animal welfare investigations.

The plasticity of the brain structures of farm animals at the neonatal period was raised in connection with neurochemical and neuroendocrinological indicators of welfare.

It is well known, that the brain of the newborn is still highly plastic and early postnatal entrainment or treatment may permanently alter or modify brain development. If this early experience is put into the context of a later response - e.g. behavioural, neurochemical or endocrine - to a stressful or other adverse situation, couldn't such treatment modify the indicators measured? It was felt that the conditions under which farm animals may be reared during the neonatal and early postnatal period need very careful definition if results are to be meaningful or comparable.

Glucocorticoids are often used to quantify acute, short-term stress situations, but we may be concerned with long-term effects. Therefore, P. Mormède/France and R. Dantzer's/France use of ACTH challenge is particularly useful. A study of the kinetics of the response to ACTH challenge during adaptation would be very valuable, and this remains to be done.

It was observed in relation to G. Beuving's/Netherlands paper that, if corticosterone is indeed an indicator of stress, and stress reduces egg production, then it should be possible to give a figure for the concentration of corticosterone at which the hen stops egg production.

SESSION II

SIGNIFICANCE OF INDICATORS RELEVANT TO ANIMAL WELFARE
Ethological indicators

Chairman: M. Zanforlin

ON THE SIGNIFICANCE OF ETHOLOGICAL CRITERIA FOR THE ASSESSMENT OF ANIMAL WELFARE

P.R. Wiepkema
Landbouwhogeschool
Marijkeweg 40
6709 PG Wageningen, The Netherlands

ABSTRACT

In this contribution the measurableness of animal welfare is discussed. For that purpose a number of abnormal behaviours in farm animals are described briefly, and classified. The evaluation of these categories of abnormal behaviour can be evaluated if one is explicit about the organisation of normal behaviour. Two models are discussed: the psychohydraulic and the regulatory one. It appears that the latter model offers the best perspectives in animal welfare research, if one wants to integrate ethological and physiological approaches. Some rules of thumb are given to evaluate ethologically the weight of Sollwerte.

INTRODUCTION

The search for biological indicators of animal welfare reflects our need to make this concept operational for those animals which we have a special responsibility: farm animals. Assuming this as our starting point, the crucial question becomes then whether or not animal welfare can be quantified like many other biological concepts. In my opinion, this goal can only be reached if we are able to formulate a biologically adequate model or representation of the way an animal normally functions.

This functioning reveals itself in ethological and physiological processes by which the animal adapts optimally to its Umwelt. Although these processes are inseparable, I shall restrict myself to the ethological ones when trying to first answer the question whether specific behaviours or classes of behaviour can be used as criteria for the assessment of animal welfare; or perhaps better, for the lack of this welfare.

Behavioural phenomena always imply the occurrence or non-occurrence of well-defined movements or postures. Some of them are called abnormal or disturbed. What are the characteristics of such behaviours and how can we develop procedures by which we prove independently that these behaviours indicate a disturbed welfare (Duncan, 1980)? Or, in medical terms, how can we show that some behaviours are analogous to injuries and scars?

First I want to mention some abnormal behaviours that are often used to indicate a disturbed welfare of farm animals. It appears that such behaviours can be classified and used as criteria for the assessment of welfare itself.

The main question about these criteria, however, concerns their biological significance. This can only be answered if we have a common picture of how animals operate physiologically, ethologically and emotionally.

ABNORMAL BEHAVIOURS

For the present purpose it is unnecessary, and even impossible, to list all the abnormal behaviours of farm animals as described by the experts in this field.

It suffices to mention briefly a limited number of abnormal behaviours observed in veal calves, tethered sows, fattening pigs and battery kept laying hens. I have chosen the most striking examples.

Veal calves kept in groups or separately in crates perform a.o.: 1) excessive licking of themselves, other calves or parts of the crate, 2) intensive sucking at other calves, 3) urine drinking, 4) tongue playing (fig. 1) and 5) sham ruminating (Van Putten and Elshof, 1982).

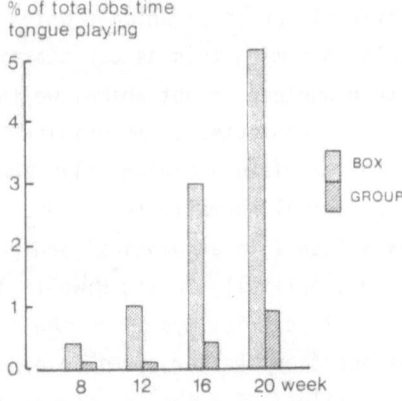

Fig. 1. The occurrence of tongue playing in veal calves kept separately in boxes or in groups of 5. The animals were bucket fed. Observations were made at 8, 12, 16 and 20 weeks of age (unpublished data of J. de Wilt, Wageningen).

Tethered sows perform a.o.: 1) bar biting, 2) rhythmic chewing and 3) motionless sitting (mourning) (Baxter, 1982; Sambraus, 1982b).

Fattening pigs perform a.o.: 1) nibbling and rooting at each other and 2) tail biting that may lead to cannibalism (Sambraus, 1982b).

Laying hens perform a.o.: 1) feather pecking that may lead to cannibalism,

2) pacing and 3) escape behaviour just before egg laying and 4) sham dustbathing (Fölsch und Vestergaard, 1981; Hughes, 1982).
This list is wilfully short in order to avoid getting lost in numerous details. But even this abbreviated list of behaviours is sufficient to show
that abnormal behaviours can be classified into a smaller number of more
general categories.

These are the following:

A. Behaviours that cause injury to the performer or conspecifics: detrimental behaviours. They may be derived from aggressive, feeding, grooming or
other behaviour systems. To this category belong the following already mentioned behaviours: urine drinking, tail biting and feather pecking.

B. Behaviours that have a constant form, are repeated over and again, have
no obvious function and find their origin in a former unsolved conflict:
stereotypes. To these belong: tongue playing, rhythmic chewing and pacing
behaviour.

C. Behaviours performed in the absence of adequate substrate or environmental
stimuli: sham behaviours (often called vacuum activities). To these belong:
rhythmic chewing, sham ruminating and sham dustbathing.

D. Some behaviours suggest or indicate a much reduced attentiveness towards
external stimuli, apathetic behaviours. To these belong: motionless sitting.

E. Behaviour strongly indicating the desire to leave the confinement in which
the animal is situated: escape behaviour. To these belong: escape of laying
hens.

F. Many behaviours are directed at an inadequate or abnormal object: redirected activities (these activities may ritualize into stereotypes). To
these belong: licking and sucking by veal calves, bar biting, nibbling and
rooting at penmates.

The point I want to make by introducing these six categories of abnormal
behaviours is that they may refer to an equal number of relevant but different aspects of animal welfare. Some behaviours may be much more severe
than others, for instance when one compares detrimental behaviours with redirected activities. Whether the evaluation of this comparison is correct
or not, we have to be aware of the possibility that not all disturbed behaviours have the same weight.

Such differences may have practical consequences when later on we want
to use the occurrence of specific abnormal behaviours as a criterion of
animal welfare in a given husbandry system. Such a criterion may have the
following form: welfare in a given husbandry system is not at stake when a

given severe abnormal behaviour does occur in less than 1% of all animals involved. For a relatively less severe abnormality one could use a 5% limit (cf. Rist, 1982).

All this means that we must clarify the biological significance of the different types of abnormal or disturbed behaviours. This can only be realized if we indicate explicitly the way we think behaviour has been organised.

MODELS OF BEHAVIOUR

In applied animal ethology two different models or schemes are used when trying to explain or to visualize the "why's" of animal behaviour. By these "why's" I mean the short term causation and function of behaviour. One model lies close to the roots of classic ethology and was designed by Konrad Lorenz (1950, 1981). In his psychohydraulic representation (fig. 2) some dammed up energy or drive is the main force behind the occurrence of specific behaviours. Animals are driven by instincts.

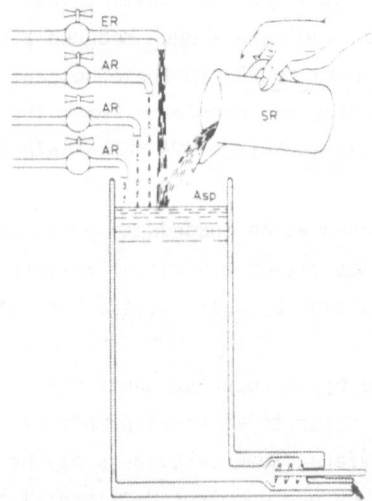

Fig. 2. In this psychohydraulic model ER represents the main source of endogenous and automatic generation of stimuli. AR represent minor unspecific readiness-releasing stimuli, and SR the effect of releasing key stimuli; Asp is level of action-specific potential. The outlet of the reservoir - behaviour - is effected by pressure within the reservoir. This pressure may be the result of ER only. The performance of behaviour reduces Asp (Lorenz, 1981).

In the second model (fig. 3), behaviour is conceived as a tool by which animals regulate their Umwelt. Homeostasis and regulation are key concepts (Wiepkema, 1981 and 1982).

The adherents of these models often use different terminology, which may lead to needless misunderstandings. Moreover both models have their own Umwelt. The Lorenzian model is in vogue in German speaking circles, whereas the regulatory or homeostatic model is typical of English speaking circles. As not belonging to both circles I am in a position to notice how few bridges there are between both ways of thinking.

In my opinion both models offer different vistas with respect to the evaluation of animal welfare and of future research in which ethology and physiology may become integrated. I hope my comments engender a fruitful discussion on the pros and cons of both approaches.

In the psychohydraulic model it is assumed that the non-performance of a behaviour brings about a damming up of some energy or drive, which at the end even may lead to the occurrence of so called vacuum activities. It is essential, that by the performance of behaviour the accumulated energy is discharged (Sambraus, 1982a). If such a catharsis is impossible the organism suffers. Such a discharge theory is also found in the older explanations of displacement behaviour (sparking-over-behaviour). The Lorenzian model strongly fosters the idea that the inability to perform a given behaviour implies a disturbed welfare, because of the resulting damming up effect. Such a relationship is not self-evident in the regulatory model.

However, the discharge idea is not absolute. Behaviour performed towards a less adequate object, for instance redirected activities, is associated with a poorer discharge (Sambraus, 1982a). Why this should be so is not explained in this model, in which feed back mechanisms are entirely absent.

Here we encounter the deficiencies of the psychohydraulic model of behaviour. Its typical terminology offers little or no points of contact for those, who strongly feel that also in the field of animal welfare research physiological and ethological approaches have to be integrated. Moreover, the idea that farm animals are driven by some blind force or instinct, may unintentionally hamper the development of cognitive ethology in our applied research. By these cognitive processes I mean that animals have and use neural representations of their Umwelt in order to control or to predict changes in that same Umwelt. This implies that animals have expectations with respect to their Umwelt and, in a sense, are aware of what is going on around them (cf. Griffin, 1982).

The regulatory model offers a good basis to integrate physiological and ethological approaches. I shall not describe the model in detail since I have done so on a previous occasion. Here I shall only mention the essential features of the model. Firstly, the driving force of a given behaviour pattern is defined by or identical to the difference between its corresponding Istwert and Sollwert. Secondly, the effect of the activities performed are fed back (negatively and positively) to the motivation involved. Thirdly, by this an organism is able to control the effectiveness of a given behaviour program.

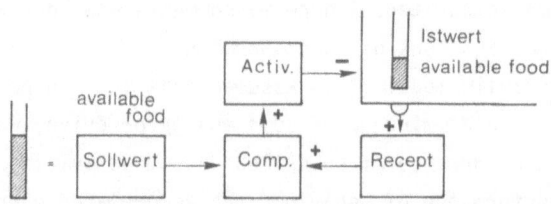

Fig. 3. Regulatory model-feeding behaviour. The Istwert of available food is perceived by an adequate Receptor and compared (Comp.) with its corresponding Sollwert. If there is a difference Activity is performed, that is negatively fed back to Istwert. Sollwert is determined by genetical, ontogenetical and experiential factors.

In regulatory systems the phenomenon of lowered thresholds and the occurrence of so called vacuum activities is explained in a way different from the one given above. Such phenomena will be observed only if an Istwert gradually moves away from its Sollwert, while the organism cannot or does not perform regulating activities. Good examples of such Istwerte are the availabilities of calories or water inside the body; both change over time as a result of metabolic processes. May sham dustbathing also be caused by a gradually changing Istwert? (cf. Vestergaard, 1982)

In the regulatory model the organism monitors the effects of its behaviour. If this does not produce the intended result, the organism may correct or change its behaviour output in that situation. In this way we may expect, that an animal in conflicting or frustrating situations initiate escape and redirected activities. Some of them may become detrimental for the animal itself or its conspecifics. Until now it is not clear why some of these conflict behaviours may ritualize and become stereotypes. It seems to occur when the original conflict is not only severe, but also unsolvable, and lasts for longer periods of time. Apathetic behaviours may be indicative of animals that stop further active coping with their Umwelt, which is over-taxing them.

The regulatory model has much overlap with present day research on stress mechanisms, in which key concepts are controllability and predictability. Both terms fit quite well in the view that organisms essentially are entities that regulate their Umwelt (cf. Levine and Ursin, 1980).

The idea that vertebrates, like man and farm animals, try to control their Umwelt, implies that they all have individual or "eigen-interests". Since all these vertebrates have been organized in a homologous way, they must experience comparable emotions, associated with being successfull or not in trying to realize their interests. I cannot see any reason why, in this respect other vertebrates should be more like machines than like men.

From this it follows that an individual's welfare is disturbed when it cannot remove by any program an existing difference between a relevant Ist-wert and its Sollwert and by this neither can control nor predict Umwelt-changes. This being unsuccessful will be associated, at least in the begin-ning, with negative emotions. We call this a form of suffering.

EVALUATION OF SOLLWERTE

It is clear that not all "eigen-interests" or Sollwerte have the same weight. Fortunately there are some rules of thumb and experimental proce-dures that enable us to assess the biological significance of a given Soll-wert.

Firstly it is plausible that irreversible Sollwerte represent more im-portant interests than reversible ones. This means that genetical and onto-genetical (imprinting) processes have more significance than those obtained by learning.

Secondly, a large difference between Istwert and Sollwert is much more burdening than a small one.

Thirdly, the weight of a given Sollwert can be estimated by measuring the amount of work (operant conditioning) the animal will perform to realize that Sollwert. In connection with this operant conditioning choice experiments may be very useful.

Fourthly, a Sollwert will be more important the more precisely an or-ganism monitors an existing difference between Istwert and Sollwert. Such a precise monitoring is demonstrated when organisms after a period of depri-vation (of food, water, sleep, a.o.) make up arrears quantitatively (cf.

fig. 4 for dustbathing in hens and Vestergaard, 1982).

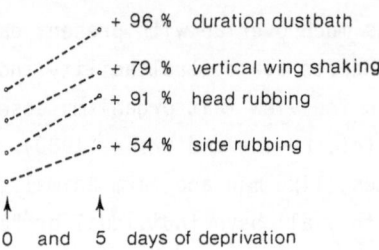

```
                          ....• + 96 %  duration dustbath
              .....•-----•  + 79 %  vertical wing shaking
      ....•-----•  + 91 %  head rubbing
  ..•-----•  + 54 %  side rubbing
 •.....
  ↑        ↑
  0   and  5   days of deprivation
```

Fig. 4. Five days of dustbath deprivation nearly doubles its duration
(increase of 96%). This increase is associated with a significant rise
of the total amounts of vertical wing shaking (+ 79%), head rubbing
(+ 91%) and side rubbing (+ 54%). Animals: 10 White Leghorns (data of
T. van Heiningen and M. de Klerk, Wageningen).

Finally, I would judge a Sollwert to be more significant the more its
non-realization is accompanied with the origin of abnormal behaviours like
the ones mentioned before.

In summary, an integration of the foregoing ethological points of view,
procedures and rules must enable us to present a scientific describtion of
behavioural injuries in farm animals. Such injuries represent real disturbances. Its description has to be combined with physiological parameters and
will lead then to a biological assessment of animal welfare.

REFERENCES

Baxter, M.R., 1982. The nesting behaviour of sows and its disturbance of
    confinement at farrowing in "Disturbed behaviour in farm animals" (Hrsg.
    W. Bessei) Hohenheimer Arbeiten 121, pp. 101-114.
Duncan, I.J.H., 1980. Animal behaviour as a guide to welfare. Feedstuffs,
    Sept., 29-39.
Fölsch, D.W. und Vestergaard, K., 1981. Das Verhalten von Hühnern (The beha-
    viour of fowl). (Birkhäuser, Basel).
Griffin, D.R. (ed.), 1982. Animal mind - human mind. Dahlem workshop Reports
    (Springer, Heidelberg).
Hughes, B.O., 1982. Feather pecking and cannibalism in domestic fowls. In
    "Disturbed behaviour in farm animals" (Hrsg. W. Bessei). Hohenheimer Ar-
    beiten 121, pp. 138-146.
Levine, S. and Ursin, H. (eds.), 1980. Coping and health. (Plenum Press, New
    York).
Lorenz, K.Z., 1950. The comparative method in studying innate behaviour
    patterns. Symposia of the Soc. for Exp. Biol. 4 (Cambridge), pp. 221-268.
Lorenz, K.Z., 1981. The foundations of ethology. (Springer-Verlag, New York),
    pp. 176-188.

Putten, G. van and Elshof, W.J., 1982. Inharmonious behaviour of veal calves. In "Disturbed behaviour in farm animals" (Hrsg. W. Bessei). Hohenheimer Arbeiten 121, pp. 61-71.

Rist, M., 1982. Beurteilungsparameter für tiergerechte Nutztierhaltungs-systeme. In "Ethologische Aussagen zur artgerechten Nutztierhaltung" (Hrsg. D.W. Fölsch und A. Nabholz) (Birkhäuser, Basel), pp. 96-108.

Sambraus, H.H., 1982a. Ethologisches Grundlagen einer tiergerechten Nutztierhaltung. In "Ethologische Aussagen zur artgerechten Nutztierhaltung" (Hrsg. D.W. Fölsch und A. Nabholz) (Birkhäuser, Basel), pp. 23-41.

Sambraus, H.H., 1982b. Sauenhaltung - wie sie ist und wie sie sein könnte. In "Ethologische Aussagen zur artgerechten Nutztierhaltung" (Hrsg. D.W. Fölsch und A. Nabholz) (Birkhäuser, Basel), pp. 49-70.

Vestergaard, K., 1982. Dust-bathing in the domestic fowl. Diurnal rhythm and dust deprivation. Applied Animal Ethology, 8, 487-495.

Wiepkema, P.R., 1981. Ein biologisches Modell von Verhaltenssystemen. In "Aktuelle Arbeiten zur artgemässen Tierhaltung 1980", KTBL-Schrift 264 (Darmstadt), pp. 15-23.

Wiepkema, P.R., 1982. On the identity and significance of disturbed behaviour in vertebrates. In "Disturbed behaviour in farm animals" (Hrsg. W. Bessei). Hohenheimer Arbeiten 121, pp. 7-17.

STEREOTYPIES AS ANIMAL WELFARE INDICATORS

D.M. Broom

Department of Zoology, University of Reading,
Reading RG6 2AJ, U.K.

ABSTRACT

Stereotyped movements form part of the normal behavioural repertoire of animals but a definition for welfare purposes is: a stereotypy is a relatively invariate sequence of movements occurring so frequently, in a particular context, that it could not be considered to form part of one of the normal functional systems of the animal. Stereotypies may sometimes be the result of malfunctions of systems controlling behaviour but they could be used to control motivational state by increasing sensory input in monotonous surroundings or by increasing the average predictability of input in unpredictable surroundings. Their frequency of occurrence should be assessed both in the normal rearing conditions and when there is human disturbance. The effects of stereotypies on the animal's body and on the pen can also be assessed. Examples of studies where rearing condition changes have reduced the incidence of stereotypies are quoted. Whatever the function of stereotypies, if they occupy much time, say 10% of waking life, the conditions could be said to be bad for the welfare of the animal. Stereotypy incidence, however, is just one of several measures which must be combined when assessing responses to adversity.

STEREOTYPIES IN RELATION TO OTHER ACTION PATTERNS

The behavioural repertoires of animals include many examples of sequences of movements which are repeated in a relatively invariate way. Such action patterns vary more or less in different circumstances and some of their components are less variable than others (Broom, 1981 p.62). For example the components of drinking by domestic chicks (Dawkins and Dawkins, 1973), display by fish (Barlow, 1977) and calling by birds (Schleidt, 1982). Action patterns are often said to be stereotyped if they do not vary much in: the number of elements in repetitions of the pattern; the degree of coupling between those elements; and the fidelity of repetition of the whole pattern including the size, duration, speed of movement and position of each component (Schleidt, 1974). Many observations are needed before such lack of variation is apparent so, in practice, sequences of movements are not referred to as stereotyped unless many repetitions have been recorded. The repetitions need not be successive for other activities may be interpolated.

Activities such as breathing, walking, drinking, or grooming may be relatively stereotyped and are not qualitatively different from the stereo-

typies sometimes shown by zoo animals, farm animals, autistic children, or
people waiting in hospitals or at bus stations.   For welfare purposes a
stereotypy may be defined as a relatively invariate sequence of movements
occurring so frequently, in a particular context, that it could not be
considered to form part of one of the normal functional systems of the
animal, feeding, body maintenance, reproduction, etc.   Some stereotypies
are incomplete (van Putten, 1982), in that they have lost elements of the
original functional action pattern but others are not distinguishable in
quality from actions which occur normally, e.g. pacing movements.

IDEAS ABOUT THE FUNCTIONS OF STEREOTYPIES

In many early studies of stereotypies it was assumed that they
represented malfunctions of systems controlling behaviour and this may be
correct in some circumstances.   Neurochemical work has demonstrated links
between the occurrence of stereotypies and dopaminergic pathways in the
brain.   For example, injection of the domamine receptor agonist
apomorphine into piglets, cattle and sheep increases the incidence of
stereotypies (Sharman and Stephens, 1974, Fry, Sharman and Stephens, 1976,
see also papers by Sharman and Oliverio in this volume).   Apomorphine
injected into rats also elicits stereotyped sequences of exploratory
behaviour (Szechtman et al., 1980).   These findings do not provide answers
concerning the function of stereotypies although the interactions between
the endorphin system and their occurrence indicate that stereotypies and
morphine-like substances may sometimes have similar effects on the brain.

Since stereotypies produce a sensory input and sensory input affects
motivational state, it is possible that animals may use stereotypies to
modify their motivational state.   Hutt and Hutt (1965) suggested that
stereotypies in autistic children "may serve to maintain arousal within
acceptable limits" and Fentress (1976) considered that stereotypy might be
a means of combatting overload.   Animals subjected to a low rate of change
of sensory input could increase input by carrying out stereotypies and,
since sensory input during stereotyped movement is very predictable,
animals in situations where events are highly unpredictable could increase
the average predictability of input by carrying out stereotypies (Broom,
1981 p.99).   If the number of operational attention channels were limited
the individual might thus reduce the necessity to process and respond to
inputs from unpleasant stimuli.   Forrester (1980) has developed the idea

that the repetition of a single behaviour pattern could be used to alter
causal factor levels and hence motivational state.   The behaviour might
be simple, as in the bill-shake of a domestic fowl (see also Hughes, 1982)
or more complex as in route-tracing stereotypies of zoo or farm animals.

SITUATIONS RESULTING IN THE OCCURRENCE OF STEREOTYPIES
      Stereotypies are often shown by animals in monotonous environments
and by those in situations of novel stimulation or high anxiety (Hinde,1970,
p.557).   They are most likely to develop in animals which have been reared
in monotonous environments and then severely disturbed.   As Meyer-Holzapfel
(1968) notes, deprivation of a specific attribute of the surroundings, such
a sleeping box or a companion, as well as monotony resulting from
restriction of movement, may elicit stereotypies.   The "high anxiety" may
result from the presence of a potential predator, such as man, or a delay
in time of feeding.   A problem for experimenters in this field is to
decide whether the occurrence of stereotypies is solely a response to the
restricted conditions in which the animal is kept or is in part a response
to the presence of the observer.
      Stereotypies shown by sows which are tethered or in pens allowing
restricted movement only include those listed in Tables 1 and 2 (Fraser,

TABLE 1.   Effects of straw on the occurrence of stereotypies in
            tethered sows.

Behaviour	% of 1 min. periods in which behaviour occurred	
	Straw	No straw
nose or lick bars, trough or chain	5	21
bite bars or chain	4	10
bite, nose or lick neighbour's tether	1	3

(data from Fraser, 1975)

1975).   It is apparent from Table 1 that tethered sows spend much time
nosing, licking or biting the accessible parts of their pens and that the
presence of straw reduces the incidence of these activities.   The sows
supplied with straw spend much time chewing and manipulating straw,
activities which are much more varied than the stereotyped licking and
biting.   The possibility that straw ingestion causes the behaviour

changes can be discounted after a further experiment by Fraser in which
straw was provided (1) on the floor and in the trough (3 kg per day),
(2) in the trough (1 kg), (3) chopped in the trough, or (4) not at all.

TABLE 2. Effects of straw presence and straw in diet on the occurrence
of stereotypies in tethered sows.

| Behaviour | % of time standing that behaviour occurred | | | |
	3 kg straw	1 kg straw in trough	1 kg chopped straw (eaten)	No straw
nose or lick bars, trough or chain	3	11	34	31
bite bars or chain	5	11	37	43
head wave	1	6	11	21
stretch mouth	0	2	4	4

(data from Fraser 1975)

Most of the chopped straw was eaten and the data in Table 2 show that the
presence of straw which can be manipulated was a major factor reducing the
incidence of stereotypies.   Similar effects of straw on the incidence of
stereotypies have been reported for individually housed calves (Unshelm,
Andreae and Smidt, 1982) but, as can be seen in Table 3, there may be
variation amongst individual calves, or amongst sets of calves, in the
stereotypy which they show so several measures are needed.

TABLE 3. Effects of straw on the occurrence of stereotypies in
individually housed calves.

| Behaviour | % of time showing behaviour | | |
	straw	no straw (1)	no straw (2)
gnawing at wood	3	5	8
tongue playing	3	6	1

(data from Unshelm, Andreae and Smidt,1982)

Both of the above examples suggest that stereotypies are shown as a
response to restricted, monotonous surroundings.   The provision of straw
increases the variety of the surroundings and makes possible a more normal
activity, i.e. chewing and manipulating the straw.   It is likely that
other kinds of added variety in the surroundings would have a similar
effect but the specific inputs obtained from something taken into the
mouth may be the most effective stimuli for reducing the incidence of
stereotypy.   At some times in life the lack of a specific stimulus or
set of stimuli may elicit stereotypies, for example the pacing shown by hens
shortly before laying if they do not have access to a suitable nest site
(Wood-Gush, 1972, Brantas, 1980, Table 4).   Stereotypies may also be shown

TABLE 4.   Effect of lack of nest access on pacing by hens: last hour
            before laying.

	Access to nest	No access
Paces	105	434

(data from Brantas, 1980)

when food is lacking or when social companions are removed.

In many studies in which the occurrence of stereotypies is reported,
they are initiated by or increased in rate or intensity by some
disturbance.   Fentress (1976, 1980) reported that disturbance by a human
observer increased rates of circling movements by caged voles Clethrionomys
and caused caged hunting dogs Lycaon to revert to the original version of
a locomotor stereotypy which included jumping over an object no longer
present.   He also reported that grooming movements by mice in unfamiliar
surroundings are more stereotyped than are those shown in the home cage.
Many observations of zoo animals and farm animals are made when the
subjects have been disturbed by human observers.   As mentioned earlier,
the presence of a human observer may result in the exaggeration of
stereotypies which are performed when not disturbed but statements about
the frequency of stereotypies should refer to their frequency in normal
rearing conditions and when there is human disturbance.

The prolonged performance of stereotypies by animals in their normal
housing conditions may be detected by observing them or by their effects.
Repeated biting movements may wear away wooden bars in the pen and rubbing
movements may produce smoothed areas on the bars or wall which are rubbed
most.   Biting may lead to measurable tooth wear and self biting may
result in mutilation.   Regular rubbing movements may cause sores on the
body.

WHAT STEREOTYPIES TELL US ABOUT WELFARE

Stereotyped movements form part of the normal behavioural
repertoire of animals but the occurrence of prolonged stereotypies
indicates that the conditions are adverse for that individual.   Whether
the stereotypy occurs as an accidental consequence of some neural
mechanism malfunction or as a means of coping with conditions by
modification of motivational state as suggested above, it is a useful
indicator that the animal is under stress (see Broom, in press).   If the
stereotypy occurred for 10% of the animal's waking life, or if it caused
bodily injury, it could be said that the conditions are bad for the

welfare of the animal. Responses to adversity may be of various kinds, however, for there are other behavioural responses such as apathy and physiological responses such as high sensitivity of the adreno-cortical system or high levels of endorphin production. Hence it is desirable for the various measures to be combined in order to detect adversity. One animal may show mainly stereotypies as a response while another may show a combination of lower levels of stereotypies with other responses.

REFERENCES

Barlow, G.W. 1977. Modal action patterns. In "How Animals Communicate" (Ed. T.A. Sebeok). (University of Indiana Press, Indianapolis). pp 98-134.

Brantas, G.C. 1980. The pre-laying behaviour of laying hens in cages with and without laying nests. In "The Laying Hen and its Environment" (Ed. R. Moss). (Martinus Nijhoff, The Hague, C.E.C symposium) pp 227-234.

Broom, D.M. 1981. "Biology of Behaviour". (Cambridge University Press, Cambridge).

Broom, D.M. in press. The stress concept and ways of assessing the effects of stress in farm animals. Appl. Anim. Ethol.

Dawkins, R. and Dawkins, M. 1973. Decisions and the uncertainty of behaviour. Behaviour, 45, 83-103.

Fentress, J.C. 1976. Dynamic boundaries of patterned behaviour: interaction and self-organisation. In "Growing Points in Ethology" (Ed. P.P.G. Bateson and R.A. Hinde). (Cambridge University Press, Cambridge). pp 135-169.

Fentress, J.C. 1980. How can behaviour be studied from a neuroethological perspective? In "Information Processing in the Nervous System" (Ed. H. Pinsker). (Raven Press, New York).

Forrester, R.C. 1980. Stereotypies and the behavioural regulation of motivational state. Appl. Anim. Ethol., 6, 386-387.

Fraser, D. 1975. The effect of straw on the behaviour of sows in tether stalls. Anim. Prod., 21, 59-68.

Fry, J.P., Sharman, D.P. and Stephens, D.B. 1976. The effect of apomorphine on oral behaviour in piglets. Br. J. Pharmacol., 56, 388p.

Hinde, R.A. 1970. "Animal Behaviour: a Synthesis of Ethology and Comparative Psychology", 2nd edn. (McGraw Hill, New York).

Hughes, B.O. 1959. Discussion of paper by W. Bessei. In "Disturbed Behaviour in Farm Animals" (Ed. W. Bessei). (Eugen Ulmer, Stuttgart, C.E.C. symposium). pp 159-160.

Hutt, C. and Hutt, S.J. 1965. Effects of environmental complexity on stereotyped behaviours of children. Anim. Behav., 13, 1-4.

Meyer-Holzapfel, M. 1968. Abnormal behavior in zoo animals. In "Abnormal Behavior in Animals" (Ed. M.W. Fox). (W.B. Saunders, Philadelphia). pp 476-503.

van Putten, G. 1982. Discussion of session III. In "Disturbed Behaviour in Farm Animals" (Ed. W. Bessei). (Eugen Ulmer, Stuttgart, C.E.C. symposium). p 129.

Schleidt, W.M. 1974. How 'fixed' is the fixed action pattern? Z. Tierpsychol., 36, 184-211.

Schleidt, W.M. 1982.   Stereotyped feature variables are essential
    constituents of behaviour patterns. Behaviour, 79, 230-238.
Sharman, D.F. and Stephens, D.B. 1974.   The effect of apomorphine on the
    behaviour of farm animals.   J. Physiol., 242, 25-27p.
Szechtman, H., Ornstein, K., Hofstein, R., Teitelbaum, P. and Golani, I.
    1980.   In "Enzymes and Neurotransmitters in Mental Disease" (Ed.
    E. Usdin, T.L. Yourkes and M.B.H. Youdim). (Wiley, New York).
Unshelm, J., Andreae, U. and Smidt, D. 1982. Behavioural and physiological
    studies on rearing calves and veal calves. In "Welfare and Husbandry
    of Calves" (Ed. J.P. Signoret). (Martinus Nijhoff, The Hague, C.E.C.
    symposium). pp 70-76.
Wood-Gush, D.G.M. 1972.   Strain differences in the response to sub-optimal
    stimuli in the fowl.   Anim. Behav., 20, 72-76.

INGESTIVE BEHAVIOUR AND WING-FLAPPING IN ASSESSING WELFARE OF LAYING HENS

H.B. Simonsen

Department of Forensic and State Veterinary Medicine
Royal Veterinary and Agricultural University
Bülowsvej 13, DK-1870 Copenhagen V, Denmark

ABSTRACT

Ingestive behaviour and wing-flapping in chickens is briefly described and consequences of deprivation of the patterns
are discussed. The welfare legislation related to the two behaviour patterns is considered and the legal differentiation
between the two behavioural patterns is discussed.

INTRODUCTION

During the latest decade legislation has included behavioural needs as parameters in the evaluation of animal welfare. The term "behavioural need" is, however, not yet clearly
defined leading to difficulties during interpretation and enforcement of legislation.

The present paper is an attempt to elucidate two behavioural patterns in laying hens as well as consequences of deprivation of the patterns in order to decide whether one or both of
the patterns are to be considered as "behavioural needs" of
the birds.

THE NORMAL BEHAVIOUR OF THE TWO PATTERNS

Ingestive behaviour including eating and drinking is described by Wennrich (1978 a). During eating behaviour the
chicken scrathes the ground and pecks and swallows the food
object. During drinking the chicken pecks against a water surface or a dew drop, raises the head and swallows the water.
Also wing-flapping is described by Wennrich (1978 b). This
pattern consists of rapid, simultaneous movements of both
wings during which the wings are stretched and flapped together
above the back of the bird.
Wing-flapping is categorized as a comfort-behaviour.

EFFECT OF FOOD AND WATER DEPRIVATION ON BEHAVIOUR AND
PHYSIOLOGY

It is generally accepted that deprivation of food and
water (and thus of normal ingestive behaviour) causes an increase in general activity of animals (Baumeister et al., 1964)
According to Wood-Gush and Guiton (1967) the thwarting of
feeding behaviour in adult hens by presentation of food under
a glass cover initially caused escape behaviour followed by an
increase in "irrelevant" grooming and sleeping behaviour which
later on decreased to the control level. Simonsen (1979) found
an increase in floor pecking activity in hens on wire and
litter floor during 48 hours of food deprivation. Changes in
the agonistic behaviour was also encountered.

Hembree et al. (1980) conducted an experiment on the effect
of force moulting on the agonistic behaviour of caged White
Leghorn hens. The experiment included a premoult period where
the hens were given 14 hr light/day and a 17% protein layer
mash, a 10 days stress period without feed and 8 hr light/day
recovery period where the hens were given ground corn and 8 hr
light/day and a post moult period with same conditions as the
premoult period. The hens were confined in colony cages
(91.3 x 71.1 cm) with 10 birds in each cage and the hens were
observed for 4, 10-min periods weekly over an 8 week period,
beginning 2 weeks prior to the start of the stress period.
Fights, pecks, threaths and avoidances were recorded for each
period to determine the number of total aggressive acts as
well as the peck order stability. Non moulted hens served as
controls. It was from the experiment concluded, that the force
moulted hens performed significant more agonistic acts per
10-min observation during the 18 day recovery period than did
hens not force-moulted.

The effect on physiology of hens by deprivation of feed
and water was demonstrated by Brake and Thaxton (1979 a), who
induced moult by removing feed up to 12 days and water up to
3 days simultaneously with a reduction of the numbers of light
hours. Body temperature increased significantly during feather
loss and renewal. Packed red cell volume and hemoglobin
increased significantly immediately upon removal of feed and

water, while plasma total calcium and inorganic phosphate decreased significantly. In a similar force moulting experiment Brake et al. (1979 b) found, that plasma levels of thyroxine initially decreased upon removal of feed, but increased above control level by the sixth day of feed withdrawal. Changes in adrenal cholesterol and total adrenal steroids were not consistent. There was, however, a trend toward increased total adrenal steroids during feed withdrawal, and an increase in adrenal cholesterol upon resumption of feeding. Also Perek and Eckstein (1959) and Ben Nathan et al. (1977) demonstrated symptoms of physiological stress during food and water deprivation.

EFFECT OF WING-FLAPPING DEPRIVATION ON PHYSIOLOGY

After a period of wing-flapping deprivation by space reduction the hens immediately show this behaviour when space is increased (Black and Hughes, 1974; Wennrich, 1975). This may lead to the suggestion, that the motivation for wing-flapping activity increases during deprivation.

Deprivation of wing-flapping activity has an effect on the physiology of the hens and a consequence of the deprivation is brittle bones in the wings. D.F. King (1965) showed that osteoporosis from lack of exercise caused Cage Layer Fatigue. Although bone weakness in battery hens in the literature is divided into Cage Layer Fatigue (CLF) and general bone brittleness the syndroms is presumed to be two stages of the same disease. They are both induced by restricting movement, show the same histological changes and are both cured by providing an opportunity for exercise. Bone brittleness cannot always be relieved by food supplements, in particular with CLF symptoms, and hens in cages also seem more sensitive to non-optimum feed mixes than hens in deep litter housing. Space for exercise nearly always eliminates CLF and results in stronger bones, whether or not the same feed is used as was used during confinement in cages (Anon. 1981).

The hens that suffer from CLF in peracute or acute form will, because of their weak bones, stand a much greater chance of breaking their bones whilst being caught and transported to

the slaughterhouse than hens that were kept in deep litter or
other floor systems. From the literature it appears that bone
weakness is a problem in spent cage layers (Rowland and Harms,
1970; Ferguson et al., 1974; Moore et al., 1977), as the condition results in broken bones and accordingly in reduced carcass value. However, only few data on the problems is available. In one Danish investigation as a mean 6.5% of the hens
had broken bones, especially humeri, after transport to
slaughter. After slaughter additional 9.6% newly broken bones
were found. The study also included wire floor hens which had
frequencies of 0.5% for both measures (Anon. 1979 a). In
another investigation the frequencies for ante mortem fractures for caged layers varied from 2.7% - 5.3% with one exception
with only 0.2%. Post mortem fractures of caged layers varied
from 1.4% - 19.0%. Pennsylvania system hens had less than 0.2%
ante- and post mortem fractures whereas deep litter hens had
less than 0.1% of each (Anon. 1979 b). Similar Nielsen (1980)
found from 0 - 7.8% ante mortem fractures of cage layers' wing-
bones as compared to 0% in birds from Pennsylvania systems.

LEGISLATION

According to the UK recommendation (1971) food and water
should not be withheld for more than 24 hours, and similar
limits are prescribed in the Swedish regulations (1974).
Denmark (1950), Finland (1971), Federal Republic Germany (1972)
and Norway (1974) have in their animal welfare acts unspecified
rules regarding the provision of food and water. According to
these it is in general an obligation to provide animals with
sufficient food and water of satisfactory quality. Other
countries may have similar rules regarding animals' provision
with food and water. According to the European convention for
the protection of animals kept for farming purposes (1976),
animals shall be housed and provided with food, water and care
appropriate to their physiological and ethological needs.

Thus it seems to be beyond any doubt that egglayers must
be able to perform ingestive behaviour with only few restrictions.

Wing-flapping behaviour is not specified in any rules and it is at present under debate whether this activity should be included under the term "ethological needs" as mentioned in the European Convention and the German 1972 act.

CONCLUSION

Evaluation of farm animal welfare is based on several factors of which behaviour and physiology are major ones. The present paper focuses on two well defined behaviour patterns in egg-layers and the effect of deprivation on the birds physiology. Deprivation of the two behaviour patterns causes changes in the birds behaviour during and after the deprivation period as well as physiological disturbances. Thus it may be concluded that both behaviour patterns are important for the birds maintainance of homeostasis and welfare in a given environment. Legislation, however, distinguish very clearly between the two patterns of which the ingestive behaviour is an obligation in most national and international legislation while wing-flapping behaviour is not mentioned.

One may ask for reasons for this difference in the welfare legislation and answers to this question probably can stimulate the debate on welfare indicators in farm animal production.

REFERENCES

Anonym, 1979 a. Intensive production and farm animal welfare. Danish Ministry of Justice. Table 19.
Anonym, 1979 b. Ibid. Document no. 31 (Bone fratures in hens).
Anonym, 1981. The physiological and ethological needs of egg-layers to perform egglaying behaviour and wing-flapping. Reviews of literature presented by Society for Veterinary Ethology to The Standing Committee of the European Convention for the Protection of Animals Kept for Farming Purposes TPA/E (81) 8.
Bareham, J.R. 1972. Effects of cages and semi-intensive deep litter pens on behaviour, adrenal response and production in two strains of laying hens. Br.Vet.J., 128, 153-163.
Baumeister, A., Hawkins, W.F. & Cromwell, R.L., 1964. Need states and activity level. Psychol.Bull., 61, 438-453.
Ben Nathan, D., Heller, E.D. & Perek, M., 1977. The effect of starvation on antibody production of chicks. Poult.Sci., 56, 1468-1471.

94

Black, A.J. & Hughes, B.O., 1974. Patterns of comfort behaviour and activity in domestic fowl: A comparison between cages and pens. Br.Vet.J., 130, 23-33.

Brake, J. & Thaxton, P., 1979 a. Physiological changes in caged layers during a forced molt. 1. Body temperature and selected blood constituents. Poult.Sci., 58, 699-706.

Brake, J., Thaxton, P. & Benton, E.H., 1979 b. Physiological changes in caged layers during a forced molt. 3. Plasma thyroxine, plasma triiodothyronine, adrenal cholesterol and total adrenal steroids. Poult.Sci., 58, 1345-1350.

Ferguson, T.M., Scott, J.T., Miller, D.H., Bradley, J.W. & Creger, G.R., 1974. Bone strength of caged layers as affected by portland cement and sodium bicarbonate. Poult. Sci., 53, 303-307.

Hembree, D.J., Adams, A.W. & Craig, J.V., 1980. Effects of force-molting by conventional and experimental light restriction methods on performance and agonistic behavior of hens. Poult.Sci., 59, 215-223.

King, D.F., 1965. Effect of cage size on cage layers fatigue. Poult.Sci., 44, 898-900.

Moore, D.J., Bradley, J.W. & Ferguson, T.M., 1977. Radius breaking strength and egg characteristics of laying hens as affected by dietary supplements and housing. Poult.Sci. 56, 189-192.

Nielsen, B., 1980. Wing bone fractures in laying hens. Danish Vet.J., 63, 981-1016.

Perek, M. & Eckstein, B., 1959. The adrenal ascorbic acid content of laying hens and the effect of ACTH on the adrenal ascorbic acid content of laying hens. Poult.Sci., 38, 996-999.

Rowland, L.O. Jr. & Harms, R.H., 1970. The effect of wire pens and cages on bone characteristics of laying hens. Poult.Sci., 49, 1223-1225.

Simonsen, H.B., 1979. Effect of feed withdrawal on behaviour and egg production in white leghorns on litter and wire. Br.Vet.J., 135, 364-369.

Wennrich, G., 1975. Untersuchungen über die Bewegungsaktivität von Haushennen (Gallus Domesticus). Arch.f.Geflügelknd., 4, 113-121.

Wennrich, G., 1978 a. Verhaltensweisen beim Fressen und Trinken. In: Nutztierethologie.Ed. H.H. Sambraus, p. 257-263. P. Parey.

Wennrich, G., 1978 b. Ibid. p. 264-268.

Wood-Gush, D.G.M. & Guiton, P.H., 1967. Studies on thwarting in the domestic fowl. Rev.Comp.Anim., 5, 1-23.

Legislation:

Codes of Recommendations for the Welfare of Livestock 1971. Code no. 3. Domestic Fowls. Article 28.

Swedish code of recommendations for the welfare of animals 1974. Management of domestic fowls. Article 3.5.3.

Danish animal welfare act 1950. Article 2.

Finnish Statutory instrument of animal welfare 1971. Article 5.

Norwegian animal welfare act 1974. Article 5, no. 1.

Federal Republic of Germany's Animal welfare act 1972.Article
    2. No. 1 and 8.
European convention for the protection of animals kept for
    farming purposes. Article 3.

BEHAVIOURAL TESTS TO QUANTIFY ADAPTATION IN DOMESTIC ANIMALS

Marina Verga & C.Carenzi

Istituto di Zootecnica-Facoltà di Medicina Veterinaria-Milan
Italy

ABSTRACT

The authors expose the reasons for their interest in a psy cho-ethological study of domestic animals in function of a comparative analysis of several species and of the use of experimental test data correlated to physiological and performance parameters,for a deeper and thorough biological knowledge of the subjects for the appraisal of their adaptive capabilities, in different environments.Examples of behavioural tests are described and the results are synthetically reported on:'openfield' test on piglets and laying hens;learning tests in heifers,laying hens and dogs;imprinting tests in turkey and pheasant poults in intensive husbandry.

INTRODUCTION

There are numerous definitions of "Ethology"(Schaffner, 1955;Lorenz,1960;Eisner & Wilson,1975;Craig,1981;Hartsock,1982) which may be summed up in:'Ethology is the scientific study of behaviour',i.e. 'what an animal does'(Lehner,1979).Obviously we must describe the characters of the subject,as well as when,how where and why this subject acts in a specific way (Nielsen,1958) with a detailed functional analysis of the different patterns of behaviour or with other methods.Most important are the 'causal factors' determining a behaviour (Toates,1980).

ETHOLOGY AND PSYCHOLOGY

Ethology is related to other sciences concerned with the study of living beings,and in particular to Biology,Zoology,Physiology and Psychology.The relation with the latter are becoming increasingly more obvious pursuant the use of experimental tecniques which unite and often blend these two sciences (Lehner,1979).The analogy is illustrated by Craig's example (1981): ethologists identify releasers eliciting particular behaviours; psychologists speak about conditioning stimuli,neural structures and conditioned responses.Ethology uses studies in the natu-

re and also experimental studies.The comparative study of behaviour,both in environments not directly controlled and controlled,on animals and humans,involves on the one hand the use of
observation tecniques for the elaboration of the ethogram,and
successively the behavioural repertoire of the subjects in different environmental situations and against different back-
grounds of infantile experiences.On the other hand,alongside
with the observations,it is essential to appraise the influence
of the specific variables existing in the external or internal
environment of each subject on the individual behavioural reac-
tions.To this effect of utmost importance are the studies carried out in laboratories or in controlled environments (and rearing farms are,to a large extent,controlled environments) so
as to single out experimental variables for every test.Most interesting to the effect of the appraisal of adaptive abilities
are the studies on the reactions elicited by experimentally con
trolled stimuli.The conditioning and the associative and mnemo-
nic abilities are examples of psychological investigations in
this sector.Behavioural parameters can be singled out from direct observation of the 'spontaneous' activity,as well as on
the recording of particular reactions of the animal response to
specific stimuli skilfully presented which constitute tests for
the evaluation of individual and collective response characteristics.

PSYCHOLOGICAL AND BEHAVIOURAL TESTS

A psychological test is an objective and standardized measurement of a behavioural sample (Anastasi,1981).Many reserves
have been expressed as to the reliability and objectivity of
the various tests used for humans.The problem is frequently the
interpretation of the results,the non absolute standardization
or the incomplete control of all the independent variables influencing the sample.In relation to domestic animals,the term

behavioural tests is more appropriate.It is possible to exert
a full biological control of the experimental subjects,and re-
cord their overt behavioural and physiological responses to
quantified stimuli,correlating them and checking the hypothesis
which had determined the choice of the tests.It is plain that
the accurate choice of the methods to be used chiefly implies:
1)the knowledge and understanding of the test to be used;2)the
basic bio-ethological knowledge of the subjects;3)the control
of the environmental variables and of homogeneity of the sample;
4)theorical reference points derived from previous experiences
described in the literature;5)accurate choice of behavioural
tests.Animal and Comparative Ethology have by now extended their
respective field of investigation beyond the laboratory struc-
tures.In fact,as Kilgour (1976) puts it 'psychologists could
make a significant contribution in a number of areas of direct
and practical relevance to farm animal behaviour'.He also recal
ls that as early as 1915 Yerkes & Coburn used pigs for psychological researches;Liddell (1921;1925) carried out investigations on sheep;Pearl Gardner (1937 a;b;1945) made observations
on horses,milk cows and sheep to study learning abilities.We al
so recall the studies on associative learning and problem-solving on equine species (and especially the famous Clever Hans
and Elberfeld Horses,Katz,1937).The use of tests may be related to biophysiological responses (Dantzer & Morméde,1979);for
instance the learning and discriminative abilities,the orienta
tion in a foreign environment,the refractiveness to response or
macroscopic stress reactions may be accompanied by hormonal,
cardiac and other alterations related in turn to the quality
of the performances (Scott & Fuller,1965;Baldwin,1977;Dewsbury,
1978;Moore et al.,1975;Wilson et al.,1975).Among the behavioural tests used for domestic animals we recall:-)positive or ad
versative conditioning and problem solving,in cattle (Kiley-

Worthington & Savage,1978;Craig,1981;McDonald et al.,1981);in equine species (Rubin et al.,1980);in pigs (Craig,1981;Parrott & Baldwin,1981);in sheep (Baldwin & Start,1981;Siegel & Moberg, 1980);in domestic fowl (Zolman & Mattingly,1981);in rabbits (Karnup & Zhadin,1980),and obviously in dogs (Rudenko & Struchov,1980;Lawicka,1979;Dumenko & Nosar,1980;Poltyreva & Petrov, 1981;Fonberg,1981).-)Discrimination of shapes,colours and other field structural characteristics in horses (Mader & Price,1981); in cattle and sheep (Baldwin,1981).-)Particular reactions to controlled sound,visual or other stimuli (Odberg,1978;Csermely & Wood-Gush,1981;Pollock & Hurnik,1978).-)Choice and preference tests to appraise space,social,feed characteristics and more generally environmental characters preferred by animals both immediately and following previous experiences (Hindman,1981; Fisher & Davis,1981;Johnston & Gottlieb,1981;Dawkins,1980;1981). -)Open-field tests to quantify exploration,emotivity and fear reactions in a new environment,on fowl (Suarez & Gallup,1980; 1981;Faure,1979;1980 a;b;Faure & Jones,1981;Faure $ Folmer,1975; Gallup & Suarez,1980);on piglets (Fraser,1974);on sheep (Winfield et al.,1981);on cattle (Kilgour,1975);on rabbits (Denenberg et al.,1981).-)Imprinting (approach and discrimination) tests to appraise the duration,temporary localization,intensity of the sensitive period and its modifications in relation to environmental and drug factors,and to the type of neonate tested (Gaioni et al.,1978;Verga & Cavalchini,1982;etc.).

EXPERIMENTAL EXAMPLES OF BEHAVIOURAL TESTS ON DOMESTIC ANIMALS

Briefly reported are some results and considerations on the use of tests for the possibility of quantification of the adaptiveness and correlation of the reactions to physiological and productive factors.

Open-field test- Subjects:18 female piglets at weaning age (29 days) divided into 3 groups of 6,of which 3 heavy and 3 light weight per group,in function of the farrowing crate which favours a more intense contact with the sow owing to the location of the heat lamp.Experimental apparatus:wooden crate with white walls and divided in squares;total surface 4 sq.mt.The results are reported in Table 1 and 2.

Table 1 - Open-field reactions in female lightest heaviest piglets among the 3 treatments. Mean values.

Female piglets	Treatm. 1		Treatm. 2		Treatm. 3	
Behaviours	H. /	L.	H. /	L.	H. /	L.
	N=3	N=3	N=3	N=3	N=3	N=3
1.Latency (sec.)	41.6± 31.7	30.6± 25.1	28.33± 27.5	3.66± 1.55	22± 32.9	21± 25.3
2.Vocaliz. (freq.)	300.0± 0	93.3± 5.7	300.0± 0	133.33± 149.77	240.0± 103.8	258.6± 69
3.Sq. ent. (freq.)	60 ± 40.8	64.6± 12.7	65. ± 20.7	81.66± 20.03	78.3± 40.4	54.6± 6.5
4.Immob. (sec.)	25 ± 30.4	53.5± 50.3	13.6± 15.1	10 ± 13.2	5.6± 5.1	5 ± 5
5.Exit (sec.)	43.3± 28.8	41. ± 32.9	41 ± 32.9	12 ± 15.5	23.3± 10.4	21.7± 17.6
6.Defecat. (freq.)	0.5± 0.5	0.6± 1.1	0.3± 0.5	1.3± 0.5	0. ± 0	0. ± 0
7.Att. to fle. (freq.)	3 ± 4.3	4.6± 6.4	1 ± 1	3.6± 5.5	4.7± 4.7	5 ± 2.6
8.Retreat. (freq.)	0 0	0.6± 0.5	0 0	0 0	0.7± 1.1	0.7± 1.1

H.=Heaviest subjects.   L.=Lightest subjects.

Table 2 - Time use by the piglets during observations (Min.).

Treatm.	Sucking	A	B	C	D	E	Total
1	61.55± 9.3	38.6± 3.6	10.5± 5.1	1.8± 1.6	6.7± 2.4	1.2± 0.9	58.44± 9.37
2	60.1 ± 9.6	51.6± 9.1	14.6± 3.9	1.2± 0.2	5.9± 1.9	1.1± 0.8	74.9 ± 9.2
3	85.9 ± 9.4	18.3± 6.3	13.2± 4.3	0.1± 0.3	0.9± 1.2	1.2± 1.5	34.01± 9.4

Legend: A=Sleep - B=Isolation - C=Interaction with mother - D=Play with littermates - E=Fight for teat order.

Differences are observed in the latency time,number of squares entered,stays and exits between the less heavy subjects and the others in Treat.2,in close proximity to the sow.The more intense activity of these piglets may indicate a stronger sense of exploration but,considering the higher number of defe cations and the lower number of exits one may presume that these animals resent to a greater extent fright stress and thus the strain of the new environment.In Treatm.3,with larger farrowing crates,both heavier and lighter piglets exhibit a high emotional level with reduced latency and numerous attempts to fleeing.In these farrowing crates,may be owing to the available space,the dams were particularly stimulating and protective of the litter until weaning.Defecations in this group were not recorded given that all the animals had defecated in the prestart container.These data are in agreement with the time use by the piglets which is an index of the greater or lesser habit with the mother's contact.

Subjects:laying hens,20 on litter and 20 caged from the start of egg-laying.The test was repeated monthly since start of egg laying for 11 months.Experimental apparatus:4 sq.mt. in wood, with floor divided in squares and white walls.The results are reported in Table 3.

Table 3 - Open-field test reactions in laying hens on litter and caged, in individual monthly observations.

Observat.	1		2		3		4		5		6		7		8		9		10		11	
Behaviours	L	C	L	C	L	C	L	C	L	C	L	C	L	C	L	C	L	C	L	C	L	C
Latency (%)	15	0	15	10	45	30	40	15	30	20	45	15	50	35	70	40	60	40	65	40	60	35
Sq. Ent. (freq.)	153	0	143	17	142	23	28	39	50	22	73	24	109	22	81	31	66	35	46	46	247	48
Defecat. (freq.)	17	14	16	10	12	9	11	10	10	6	6	5	11	12	21	8	6	8	6	6	9	6
Att.to Flee. (freq.)	5	4	6	0	4	1	2	1	0	2	0	0	1	1	6	0	2	3	4	0	0	4
Exit (%)	25	0	20	5	50	15	25	15	10	5	20	5	40	5	30	0	5	5	35	5	30	10

Laying hens C defecated ever before start.

Considerable differences are recorded in the percentages of spontaneous entries,number of squares entered,defecation frequencies and attempts to fleeing in the tests at different ages.In hens on litter the behaviour probably reveals an habit to the test environment alongside with an increase in the number of spontaneous exits,dicrease and stabilization of locomotor activity associated with defecation and fleeing.In cag ed hens there is a strong "freezing"reaction that,associated with the other parameters considered,may suggest a stronger fear stress and a disadaptive reaction.These behavioural data

are associated with significant differences in production and
metabolism.The different reactions detected in 'open-field'
may be assigned to stronger fear reactions in caged hens,less
adaptable to the new environment;but this might be also the
result of conditioning and habit to the absence of any locomotory activity,so that these animals would not have learnt to
move owing to objective space limitation,and thus would exhibit
only the behaviours known to them and not indicative of a higher or lower stress.

Learning tests- Subjects:90 Friesian cows,free stabled.
Motivation:feed drive.Positive reinforcement:obtainment of
concentrate.The hypothesis formulated was the correlation betwe
en learning ability (operant conditioning) and:social hierarchy,
activity levels,feeding times,in consideration to their relation
to the individual adaptive ability.

    After the establishment and stabilization of the hierarchy,
the presumed reactions were checked,and showed that both learning was rapidly consolidated in cows,and dominance,correlated
to the levels of activity and feeding times,makes possible a
better adaptation to the environmental structures;presumably
the subjects with better learning ability feed themselves first,
dominate the other cows and obtain a series of advantages.

    Subjects:20 laying hens on litter and 20 caged.
Experimental apparatus:maze divided in two parts by a dark,dull
wooden inclined structure on which the henscould perch.
Motivation:feed drive.Positive reinforcement:obtainment of feed.
The hypothesis formulated was the higher capacity of feed de-
tection in the animals used to move freely.The results are reported in Table 4.The hypothesis was checked by the percentage
of subjects reaching the feed and by the number of trials and
the time employed to reach the feed.

Table 4 - Performance in the attaining-food test after different times of feed deprivation
in laying hens reared on litter (L) or caged (C) from the beginning of laying.

Test	L:N=20			C:N=20		
	I (24 h.)	II (30 h.)	III (30 h.) (displaced feed)	I (24 h.)	II (30 h.)	III (30 h.) (displaced feed)
Reached	60%	85%	85%	10%	5%	25%
Not reac.	40%	15%	15%	90%	95%	75%
Time (×)	199.5± 142.2	101.4± 137.4	105.8± 129.8	332.5± 102.5	60"	376"
N.Attem.	5.05± 4.47	1.85± 1.69	2.55± 1.36	1.55± 2.66	1.8± 3.76	1.25

Subjects:22 dogs (Rottweiler),11 reared in families since the age of 2 months,11 always reared in kennels.Test age:8 mon.

Motivation:feed drive versus self-defence.

Experimental apparatus:meat dish (tested for appetibility with all the subjects) to be discovered behind a wooden tunnel under which the dogs had to pass to obtain the reinforcement.

The hypothesis formulated was the greater ability of finding the feed,and a lower number of avoidance (fear) reactions in the subjects with a larger number of previous experiences in a richer environment than the kennel.

Results showed that animals reared in families score better explorative and problem-solving performance than the others.

In fact 82% of dogs in the first group reached the meat (av.time 4'30"),and only 63% reached meat in the second group (av.time 6'30").

Imprinting tests- Subjects:40 turkey poults,divided in 4 experimental groups per imprinting stimulation (an intermittent electronic sound stimulus imitating the hen's call):only before hatching;only after hatching;before and after hatching;control.

It was hypothized a stronger approach and discrimination reaction to the imprinting object in the stimulated subjects, and possible differences in function of the stimulation time.

The results are shown in Figure 1.

Subjects:44 pheasant poults divided in 4 groups as above.

Imprinting stimulus:intermittent electronic sound stimulus imitating the hen's call,combined with an intermittent green dim
light stimulus.The results are shown in Figure 2.

Figure 1 -Individual approach (A) and discrimination test (B)
         in turkeys poults.

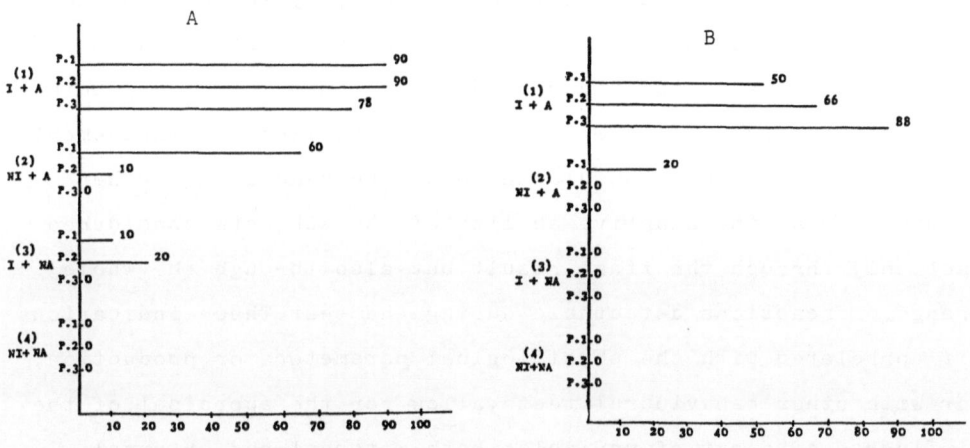

Figure 2 -Individual approach (A) and discrimination test (B)
         in pheasant poults.

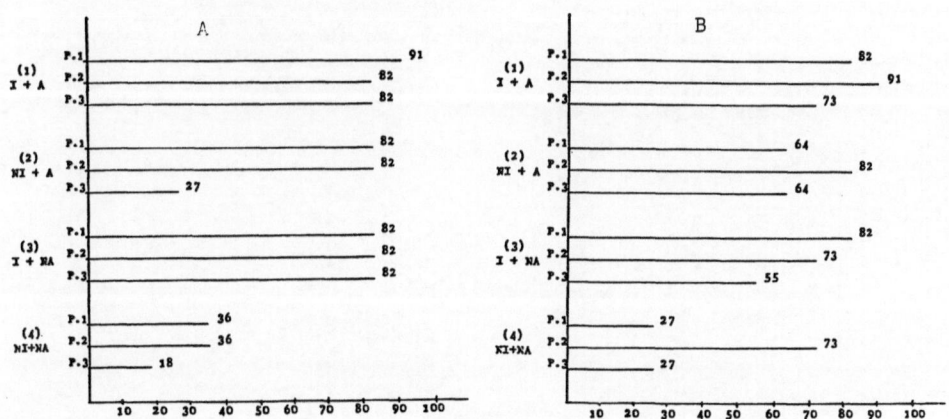

Legenda:I+A=stimulus before and after hatching;NI+A=only after
hatching;I+NA=only before hatching;NI+NA=control group.
P.1=24 h.(age);P.2=72 h.(age);P.3=144 h.(age).

   The statystical analysis of the data revealed significant
differences among the groups in the percentages of subjects
and in the time of reaching of the imprinting stimulus.

CONCLUSIVE REMARKS

The tests described above,whose preliminary results are schematically reported,don't obviously provide final conclusions on the choice nor on the methods to be used for an accurate evaluation of adaptivity.However one may infer from their application that the evaluation of behavioural reactions may contribute to a better understanding of domestic animals both of their characters and in relation to handling and management.

The test results should therefore be read as a further indication of the adaptive ability of the subjects considered not only through the final result,but also through the whole range of reactions detectable during the test.These indications if correlated with the physiological parameters or production, or with other behavioural tests,allow for the appraisal of the influence of a set of variables,both internal and external, for an improved animal-man-artificial habitat relationship,and for a more accurate quantification of the adaptation.

REFERENCES

Anastasi A.,1981,Franco Angeli (Ed.),Milano,6th Edition.
Baldwin B.A.,1981,Anim.Behav.,29,830-834.
Baldwin B.A. & Meese G.B.,1977,Anim.Behav.,25,497-507.
Baldwin B.A. & Start I.B.,1981,Proc.R.Soc.Lond.Ser.B.,211,1185,
    513-516.
Cherfas J.J. & Scott A.,1981,Anim.Behav.,29,301.
Craig J.V.,1981,Domestic Animal Behaviour,Prentice Hall,Inc.,
    Englewood Cliffs,N.J.
Csermely D. & Wood-Gush D.G.M.,1981,Biol.Behav.,6,159-165.
Dantzer R. & Morméde P.,1979,Le stress en èlèvage intensif.IN-
    RA & Masson,Paris.
Dawkins M.,1980,Anim.Regul.Stud.,3(1-2),57-63.
Denenberg V.H.;Zeidner L.;Rosen G.D.;Hofmann M.;Garbanati J.A.;
    Sherman G.F. & Yutzey D.A.,1981,Dev.Brain Res.,1(2),165-
    169.
Dewsbury D.A.,1978,Comparative Animal Behaviour,McGraw Hill
    Book Company,New York.
Dumenko V.N. & Nosar V.I.,1980,Zh.Vyssh.Nervyn.Deyat.I.P.Pavlo-
    va,30(5),927-936 (Abstr.).
Eiserer L.A.,1978,Anim.Learn.Behav.,6(1),27-29.
Eiserer L.A.,1980,Exp.Anal.Behav.,33(3),383-395.
Eisner T. & Wilson E.O.,1975,Animal Behaviour:Readings from
    Scientific American,W.H.Freeman & Co.,San Francisco.
Faure J.M.,1979,Biol.Behav.,4,241-248.
Faure J.M.,1980,Biol.Behav.,5,29-35.
Faure J.M.,1980,Appl.Anim.Ethol.,6,385.
Faure J.M. & Folmer J.C.,1975,Ann.Génét.Sél.Anim.,7(1),123-132.
Faure J.M. & Jones R.B.,1981,IRCS Med.Sci.,9,112-113.
Fonberg D.;Kostarzyk E. & Prechtl J.,1981,Pavl.J.Biol.Sci.,16,
    4,183-193.
Fox M.W.,1968,Abnormal Behaviour in Animals,Saunders (Ed.),
    Philadelphia.
Fisher G.J. & Davis S.J.,1981,Dev.Psychobiol.,14,3,237-249.
Fraser D.,1974,Applied Anim.Ethol.,1,3-16.
Gaioni S.J.;DePaulo P. & Hoffman,H.S.,1980,Anim.Learn.Behav.,
    8(4),673-678.
Gallup G.G. & Suarez D.S.,1980,Anim.Behav.,28,368-378.
Hartsock T.G.,1982,J.Anim.Sci.,54,2,447-449.
Hemsworth P.H.;Brand A. & Willems P.,1981,Liv.Prod.Sci.,8,67-74.
Hindman J.L.,1981,Dev.Psychobiol.,14,1,13-18.
Johnston T.D. & Gottlieb G.,1981,J.Comp.Physiol.Psychol.,95,5,
    663-675.
Karnup S.V. & Zhadin M.N.,1980,Zh.Vyssh.Nervyn.Deyat.I.P.Pavlo-
    va,30 (5),964-970 (Abstr.).
Katz D.,1937,Animal and Men,Longman Green,London.
Kiley-Worthington M. & Savage P.,1978,Applied Anim.Ethol.,4,119
    124.

Kilgour R.,1975,Anim.Behav.,23,615-624.

Kilgour R.,1976,Applied Anim.Ethol.,2,197-205.

Lawicka W.,1979,Acta Neurobiol.Exp.,39 (6),537-552.

Lehner P.N.,1979,Handbook of Ethological Methods,Garland Press,
    New York and London.

Liddell H.S.,1921,Proc.Soc.Exp.Biol.Med.,19,423-425.

Liddell H.S.,1925,Am.J.Psychol.,36,544-552.

Lorenz K.,1960,Methods of approach to the problems of behaviour.
    In:The Harvey Lectures 1958-1959,Academic Press,New York.

Mader D.R. & Price E.O.,1980,J.Anim.Sci.,50,5,962-965.

McDonald C.L.;Beilhartz R.G.& McCutchan J.C.,1981,Applied Anim.
    Ethol.,7,113-121.

Moore C.L.;Whittlestone W.G.;Mullord M.;Priest P.N.;Kilgour R.
    & Albright J.L.,1975,J.Dairy Sci.,58,1531-1535.

Nielsen E.T.,1958,Proc.10th Int.Congr.Entomol.,2,563-565.

Odberg F.O.,1978,Equine Vet.J.,10(2),82-84.

Parrott R.F. & Baldwin B.A.,1981,Physiol.Behav.,26,3,419-422.

Pearl Gardner L.,1937,J.Comp.Physiol.Psychol.,2,305-332.

Pearl Gardner L.,1937,J.Comp.Psychol.,23,35-57.

Pearl Gardner L.,1945,J.Comp.Psychol.,38,343-357.

Pollock W.E. & Hurnik J.F.,1978,J,Dairy Sci.,61,11,1624-1626.

Poltyreva T.E. & Petrov E.S.,1981,Zh.Vyssh.Nervyn.Deyat.I.P.Pa-
    vlova,31,3,472-478 (Abstr.).

Robinson-Guy E.D. & Schulman A.H.,1980,Behav.Processes,5(3),
    211-225.

Rubin L.;Oppegard C. & Hintz H.F.,1980,J.Anim.Sci.,50,6,1184-
    1187.

Rudenko L.P. & Struchov M.I.,1980,Zh.Vyssh.Nervyn.Deyat.I.P.Pa-
    vlova,30(2),265-271 (Abstr.).

Schaffner B.(Ed.),1955,Group processes:transaction of the Ist
    Conf.,Josiah Macy,Jr. Found.,New York.

Scott J.P. & Fuller J.L.,1965,Genetics and the social behaviour
    of the dog.University of Chicago Press.

Suarez S.D. & Gallup G.G.,Jr.,1980,Bird Behav.,2(2),93-105.

Suarez S.D. & Gallup G.G. Jr.,1981,Anim.Learn.Behav.,9,2,153-
    163.

Toates F.,1980,Animal Behaviour,J.Wiley & Sons Ltd.,Chichester.

Verga M. & Cavalchini G.L.,1982,Rivista di Avicultura,9,61-68.

Wilson J.C.;Albright J.L.;Collins J.L.;Bugden G.;Eden A. &
    Buesnel R.J.,1975,J.Dairy Sci.,58,749.

Winfield C.G.;Syme G.J. & Pearson A.J.,1981,Applied Anim.Ethol.,
    7,67-75.

Wood-Gush D.G.M. & Csermely D.,1981,Anim.Prod.,33,107-110.

Yerkes R.M. & Coburn C.A.,1915,J.Anim.Behav.,5,185-225.

Zolman J.F. & Mattingly B.A.,1981,Anim.Behav. Learn.,9,2,178-
    182.

EFFECT OF SOCIAL AND ENVIRONMENTAL FACTORS
ON PERFORMANCE IN GILTS, BARROWS AND BOARS

Anne Mette Hagelsø   Laurits Lydehøj Hansen
National Institute of Animal Science
Department for Research in Pigs and Horses
Rolighedsvej 25, 1958 Copenhagen V, Denmark

ABSTRACT

As pig rearing systems become increasingly intensive, the effect of
restriction is apparent in both the behaviour and the productive performance of the animals. A frustrating competitive environment often leads to a
high level of aggressive behaviour and a stressful condition for most of the
animals in the group. As a result, reduced feed intake and feed utilization
will lead to poorer growth rate. The three sexes have different potentials
for growth, which apparently are influenced by both feeding regime and the
type of social interaction within a group. Social rank influences the
amount of taint in boar fat, and other environmental factors also seem to
play a part. Stressful experiences during transport and preslaughter increase the incidence of carcasses with poor meat quality. Differences between breeds are probable in almost all the described conditions.

Before the days of impossibly narrow profit-margins and resulting large,
one-manned intensive rearing systems, the farmer had no need to make detailed calculations on the relationship between what was put into the animal
and what it gave in return. There was room for a deal of variation and a-
daptation at all stages. Nowadays the farmer seems obliged to choose an
often rigid system enabling him to produce most effectively and in which
the productive potential of the single animal must give way to the overall
remunerativeness of his production unit.

The main performance traits in pig production are growth rate, feed efficiency, carcass quality (meat content) and meat quality (tenderness, water binding capacity, hedonic qualities).

These traits are under the influence of other than the quantity and quality of nutrients fed to the animal and these other factors become increasingly important under intensive management conditions.

We have examined the function of the social hierarchy within pens of
pigs under different conditions. If ad lib. feeding is to have advantage
over restricted feeding methods, it must be so practised that competition
for both feed and feeding space is minimized.

Experiments were carried out with pigs having access to either one or several feeders per pen (Hansen et al. 1982). It was shown that in pens with one self-feeder, a competitive situation arose due to the socially facilitated feeding behaviour by which all pigs are prompted to eat at once. The dominating pigs in the hierarchy benefitted from this at the expense of the lower-ranking ones, and the pigs showed aggression, eating activity and weight gain according to rank. In pens with several feeders, eating activity and weight gain were independent of rank and the intra pen variance in weight gain was less than in pens with one feeder. The average amount of eating activity per pig was significantly higher and moreover, the frequency of aggression, tailbiting and earbiting was significantly lower than in pens with one one feeder, demonstrating that competition for food can result in a great deal of injurious behaviour.

Behavioural registration took place at three stages – in the beginning, the middle and the end of the growing period. In the middle period of both groups, gilts proved significantly more aggressive than barrows: in pens with several feeders due to the barrows' much reduced aggressiveness but in pens with one feeder due to gilts' much increased aggressiveness. Without being able to explain the physiological background for this, we conclude that gilts under restricted, ie. frustrating conditions are much more aggressive than barrows.

The work of Hansen (1977) supports this conclusion in that gilts, in pens with wholly slatted floors and no bedding, proved to be tailbiters in 75% of the registered cases, which took place mainly in the middle of the growing period (ie. approx. 50 kg liveweight).

Others have shown that competition for water also can be the cause of both increased aggression and excess water consumption (Olsson, 1982).

We used no straw bedding in the above mentioned experiment, but sawdust. Based on experiments (review by Hansen & Hagelsø, 1980) and experience, we are of the opinion that straw bedding has a beneficial effect in an otherwise deficient environment. Not only is it comfortable bedding and has occupational value for the pig, but more important, it seems to have special frustration-reducing properties not present in other types of bedding material or "toys". Straw is edible, hereby satisfying the pigs' need for foodseeking (foraging)-behaviour and this might explain why tailbiting seldom is seen amongst pigs supplied with straw bedding.

We have completed the practical section of a series of four 2^3 factorial experiments to investigate the effects of rearing pigs in daylight or complete darkness, with or without straw bedding and at two different space allotments. All pigs have been immunized and blood samples taken in the course of the experiment, to examine whether the rearing methods described affect the immune status of the pigs. Social rank has been determined and a 24 hr video recording made of the behaviour in all pens. Ad lib. feeding was practised from a feeder allowing 2, sometimes 3, pigs out of a group of 10 to feed simultaneously.

We cannot bring results from this project, as analysis of this large material is far from completed. We have a general impression though, that pigs kept in darkness are not less active, or less aggressive than those kept in daylight. Activity and resting periods do not seem to be as well defined and the pigs in darkness seem to be easier aroused by external stimuli. Time will show whether these assumptions are correct.

To be frank, we had expected tailbiting in these experiments, but hardly any occurred so we will be unable to cast further light over this problem just yet. The absence of tailbiting could be due to the fact that hybrid (Yorkshire x Landrace) pigs were used. Experience shows that breed differences do exist in both aggressiveness and the tendency to bite tails.

At the Danish progeny testing stations, pigs were previously kept individually and fed restrictedly. Under these conditions gilts gained weight slightly faster than barrows, demonstrating a better feed conversion rate. This was confirmed by Staun (1971) who also showed that boars, reared individually and fed restrictedly, both grew and converted feed significantly better than barrows.

When the stations in 1972 changed to keeping the pigs in groups comprising 2 barrows + 2 gilts and at the same time switched to ad lib. feeding, barrows grew significantly faster than gilts (60 g daily) (Nørtoft Thomsen & Pedersen, 1974).

Since 1981 pigs are reared and fed ad lib. in groups of 3: one boar, one barrow and one gilt. Based on results from 48 such groups, barrows surprisingly grow much faster than boars and boars in turn only insignificantly more than gilts (Jørgensen, 1981). Barrows are of course significantly fatter than boars, but gilts not fatter than boars.

Social interactions, feeding behaviour and appetite in the three sexes should be investigated under different feeding regimes. Further, analysis of the much larger material now available will show whether there are breed differences involved as well.

Superimposed on a long-term selection experiment for strong and weak boar taint, a project is in progress to investigate the effect of social rank on boar taint, measured by the level of 5 α-androstenone in fat tissue. This pheromone has been identified as the substance mainly responsible for boar taint (Patterson, 1968), and it is highly correlated with the level of testosterone in plasma. In addition the boars and gilts are kept separately in adjacent pens with or without straw bedding. The pigs are fed twice daily at a very high level (Jonsson et al., 1982).

Results until now show that the rank a pig achieves is partly dependent on its size at penning, more so amongst boars than amongst gilts and within each sex, more so in pens with straw bedding than without. This means that in pens without bedding other factors to a larger degree determine which rank the pigs attain, eg. a more slippery floor or aggressive behaviour.

The effect of rank on daily gain tends to be greater amongst gilts than boars and tends to be greater in pens without straw than with straw. Boars under all conditions grow faster than gilts and in both sexes pigs with straw bedding grow faster than those without.

Social rank has proved to have significant effect on the level of 5 α-androstenone in boar fat. Until now there is a tendency for rank to have greater effect in pens without straw. The mean levels of 5 α-androstenone do not differ, but if the variation can be lessened the frequency of strongly tainted animals can be reduced.

Another substance responsible for taint and claimed to act synergistically with 5 α-androstenone, is skatole. This is formed in the intestine as a putrefaction product of tryptophan (Hansson et al., 1980), but the level of taint does not seem to be determined by the amount of tryptophan in the feed (Bonneau & Desmoulin, 1981).

We plan to investigate whether the amount of skatole formed and deposited in fat can be influenced by more or less stressful environments. Serotonin is also a metabolite of tryptophan and it has in the rat been shown to play a part in the regulation of gastric secretion along with other catecholamines (Fjalland, 1973). Whether this has any effect on the amount of skatole produced, remains to be seen.

Differences between sexes in deposition of skatole in fat may be due to the differences in degree of unsaturation of fatty tissue (Malmfors et al., 1978).

A colleague at our institute conducted an experiment which aimed at improving the utilization of space, combined with reducing the intra-pen variance in growth, by increasing the stocking rate in the beginning of the fattening period (Nielsen, 1982). At approx. 50 kg liveweight the two smallest and the two largest from each pen of twelve pigs were removed and regrouped with respectively the smallest and largest of other pens, giving pens with 8 pigs of more uniform size. This resulted in a 25% increase in space utilization and a much reduced variation in weight gain within pens, allowing pens to be emptied faster.

The post-regrouping growth curves of these pigs were interesting. The small pigs gained weight faster than non-moved controls whereas the largest pigs had reduced growth rate compared to controls when fed restrictedly.

Hahn (1982) explains this through the principle of equifinality by which all organisms strive towards an endpoint in size which is genetically determined. This process is often disrupted by environmental stressors, but within certain limits of both degree and duration the organism is capable of recovering by compensatory performance after the stressor has been removed.

Based on our own work and that of others we see aggressive behaviour as an often occurring response to conflict and frustration. Animals exposed to aggression are exposed to stress and the response will be both acute and chronic when pigs are confined in a pen with a high level of aggressive behaviour. The animals at the bottom of the hierarchy will not only get too little to eat, they will also have a poorer feed conversion rate (Hemsworth et al., 1981). This will account for many of the socalled runts - pigs that are not clinically ill, but gain weight only slowly and for no apparent reason don't thrive.

In Sweden a new neuroleptic drug (as yet unclassified), with very specific anti-stress and anti-aggressive properties, has been developed (Björk et al., 1982). It has been extensively tested on pigs and proves to have long lasting positive effects on both growth and feed conversion rate after a single administration when pigs are mixed and moved into a fattening unit. Its marked effect on especially runts also supports the idea that this wasting syndrome often is due to stress.

Once the mode of action of this compound has been clarified, and it proves to have the claimed specific properties, it may be a useful tool in investigations on stress, frustration and aggression.

The effect of the experimental treatment on meat quality has most often been measured by pH in meat 45 mins. after slaughter. Hansen (1977) demonstrated a significant connection between social rank and meat quality, but since then no clear results to this effect have been brought to light.

It is apparent from a review by Grandin (1980) and from Frøystein et al. (1981) that a great many factors are involved eg. hereditary disposition, management, transport methods and duration, time since last feed, fighting underway to or at the abattoir, method of stunning and time from stunning till the pig is bled. The most recent events, especially fighting and herding with an electric prodder will mask the effects of earlier experiences, although the latter possibly predisposes the pig for a mild or severe stress response at slaughter.

If the effects of management or experimental treatment ie. long-term stressors are to be clarified, transport and slaughter must take place under very controlled conditions. As yet, this has unfortunately not been possible in our experiments. However, there can be no doubt that animals for slaughter should be subjected to as few stressful experiences as possible if the frequency of PSE and DFD carcasses is to be minimized.

REFERENCES

Björk, A. et al. 1982. The clinical effects of Amperozide - a novel anti-stress compound - in pig production. 3 Abstracts in: Proceedings I.V.P.S. Congress, Mexico. 315-317.

Bonneau, M. & Desmoulin, B. 1981. Influence de l'exces de tryptophane et des conditions d'elevage sur la frequence des odeurs sexuelles des jeunes porcs males entiers: relation avec le developpement de l'appareil genital. Journees Rech. Porcine en France. 329-334.

Fjalland, B. 1973. Adrenergic and Seotinergic Mechanisms in Gastric Secretion in Rats. Acta Pharmacol. et toxicol. 33: 103-112.

Frøystein, T., Slinde, E. & Standal, N. 1981 (editors). Proceedings from Symposium on Porcine Stress and Meat Quality. Agricultural Food Research Society, Norway.

Grandin, T. 1980. The effect of stress on livestock and meat quality prior and during slaughter. Int. J. Stud. Anim. Prob. 1: 313-337.

Hahn, G.L. 1982. Compensatory performance in livestock: influences on environmental criteria. Proceedings of the 2nd Int. Livestock Environment Symposium (Publ. Am. Soc. Agric. Eng.) 285-294.

Hansen, L.L. 1977. The stability of the dominance hierarchy in growing pigs in different environments. C.I.G.R. section II. Seminar on Agricultural Buildings. Band 1. 179-187.

Hansen, L.L. & Hagelsø, A.M. 1980. A general survey of environmental influence on social hierarchy function in pigs. Acta Agric.scand. 30: 388-392.

Hansen, L.L., Hagelsø, A.M. & Madsen, A. 1982. Behavioural results and performance of bacon pigs fed ad libitum from one or several selffeeders. App. Anim. Ethol. 8: 307-333.

Hansson, K.-E., Lundström, K., Fjelkner-Modig, S. & Persson, J. 1980. The Importance of Androstenone and Skatole for Boar Taint. Swedish J. Agric. Res. 10: 167-173.

Hemsworth, P.H., Barnett, J.L. & Hansen, C. 1981. The Influence of Handling by Humans on the Behaviour, Growth and Corticosteroids in the Juvenile Female Pig. Hormones and Behaviour 15: 396-403.

Jonsson, P., Hagelsø, A.M., Jørgensen, J.N. & Bach, E. 1982. The relationship between social ranking and other measurable traits in entire male pigs. 33rd EAAP meeting. Leningrad. (Under revision).

Jørgensen, B. 1981. Unpublished data.

Malmfors, B., Lundström, K. & Hansson, I. 1978. Fatty acid composition of porcine backfat and muscle lipids as affected by sex, weight and anatomical location. Swedish J. Agr. Res. 8: 25-38.

Nielsen, E.K. 1982. Increased stocking rate combined with less variation in slaughter weight amongst fattening pigs. (Danish). 435th publ. Nat. Inst. Anim. Sci. Copenhagen. 4 pp.

Nørtoft-Thomsen, R. & Pedersen, O.K. 1974. Sammenlignende forsøg med svin. 417th Report. Nat. Inst. Anim. Sci. Copenhagen. 45 pp.

Olsson, O. 1982. Evaluation of bite drinkers for fattening pigs. Proceedings of 2nd Int. Livestock Environment Symposium (Publ. Am. Soc. Agric. Eng.) 225-233.

Patterson, R.L.S. 1968. 5 α-Androst-16-en-3-one, compound responsible for taint in boar fat. J. Sci. Food Agric. 19: 31-38.

Staun, H. 1971. Experiments with boars and different methods of castration. Yearbook: Royal Vet. Agric. Univ. Copenhagen. 60-71.

OPERANT CONDITIONING IN FARM ANIMALS AND ITS RELEVANCE TO WELFARE

B.A. Baldwin

Agricultural Research Council, Institute of Animal Physiology, Babraham, Cambridge, England

ABSTRACT

Operant conditioning methods, in which animals learn to alter some aspect of their physical environment, can be used to study environmental preferences in farm animals. Operant techniques have been used to study temperature and illumination preferences in pigs and ruminants and to investigate the ability of sheep and calves to discriminate between similar shapes.

INTRODUCTION

The environment of a farm animal may be considered to consist of three main components which are indicated below:

1. The Physical environment - temperature, illumination, type of floor, humidity, ventilation etc.

2. The Social environment - size and composition of the group, isolation, dominance relations etc.

3. The Managerial environment - diet, feeding routine, cleaning systems, weaning procedures.

Optimum welfare in intensively-housed livestock depends upon providing environmental conditions which cause minimum discomfort to the animals and operant methods enable us to 'ask the animals' which environments they prefer. Factors in the physical environment such as illumination and temperature can be studied by means of preference tests and measurement of the motivation to obtain a particular envieonmental situation.

When considering the results of preference tests it is important to realize that the experimenter is selecting and limiting the choices offered to the animal. For example, it is probable that an animal, given a choice between two types of floor, chooses the less aversive material and a series of tests with a wide range of materials is necessary to find the 'ideal' floor (Hughes, 1976). It has been emphasized by Dawkins(1980) and Duncan (1978) that animals may not always choose conditions which are best for their long-term health and well-being. However, despite the considerations mentioned above, which are more fully discussed by Dawkins (1977, 1980), the use of operant methods and preference tests can be very useful in evaluating welfare problems.

In our laboratory we have used operant methods to study environmental preferences for light and temperature in pigs and ruminants (Baldwin, 1979). Before we consider specific experiments it would be helpful to outline briefly some principles of operant conditioning.

Operant conditioning is a procedure which will reliably increase the frequency of occurrence of any behaviour in the repertoire of an animal. In general terms, an animal's behaviour may cause the appearance of an additional stimulus present in the environment. If the appearance of a stimulus as a consequence of a behavioural act (response) produces an increased probability of that act being repeated, the stimulus is a positive reinforcer and the process is termed positive reinforcement. If the removal of a stimulus as a consequence of a response results in an increased probability of that response occurring in the future, the stimulus is termed an aversive stimulus or negative reinforcer and the process is called negative reinforcement. In everyday language, animals will learn to respond to obtain something they want or to remove something they dislike.

ILLUMINATION PREFERENCES IN RUMINANTS AND PIGS

The level of illumination provided for animals kept in intensive husbandry systems is a matter of controversy and calves and pigs are often kept in semi-darkness in the belief that they will be more calm and prod-uctive. In order to determine the amount of lighting which animals prefer, a series of experiments has been carried out using operant conditioning techniques.

A light-proof pen has been constructed in which the illumination preferences of individual calves may be determined. The calf lives in the pen for several days during the experiment. At the end of the pen are 2 slits behind which are invisible infra-red beams. If the calf interrupts one of the beams by inserting its muzzle into the slit the pen lights are turned on, while if the calf interrupts the other beam they are turned off. Calves soon learn to operate their on/off switch and can select the durat-ion of illumination which they prefer. In this situation the calves had the lights on for an average of 67% of each 24h. However, the calves did not turn the lights off at night and there was no obvious circadian pattern of illumination (Baldwin & Start, 1981). Sheep (Baldwin & Start, 1981) and pigs (Baldwin & Meese, 1977) have been tested in similar experiments and sheep had lights on for 77% of each 24h and pigs for 72%. It is of interest

that all three species selected an approximately similar duration of
illumination.

The motivation of sheep and calves to obtain light has been examined
by means of an operant task in which they were rewarded with only 40 sec
of light for each interruption of the infra-red beam.

Sheep obtained an average of 1.5h of light per 24h and calves 1.0h per
24h. Although sheep and calves have a strong preference for light over
darkness they are not as highly motivated to 'work' for light as they are
for heat.   Recent experiments (Baldwin & Start - unpublished) have shown
that pigs in darkness will also 'work' to obtain 40 sec periods of illumination but when placed in continuous light they would not respond to
obtain 40 sec periods of darkness. It is apparent that the pigs responded
for light and not merely for a change of stimulus.

Sheep and calves have also been trained, using an operant conditioning method in which they pressed panels with their muzzles in order to
obtain food, to discriminate between simultaneously presented shapes. In
the discrimination task the animals faced two response panels upon which
were projected the shapes to be discriminated. Only presses on the panel
associated with the correct shape were reinforced, and after each reinforcement the position of the positive stimulus was randomly varied. The results
obtained demonstrated that sheep and calves can discriminate between a
variety of similar shapes (Baldwin 1981). The results of the above experiments combined with observation of their visual acuity under field conditions indicate that vision is a major sense in these species and that
housing them in semi-darkness would deprive them of a significant proportion of their sensory input.

BEHAVIOURAL THERMOREGULATION IN PIGS AND SHEEP
In these experiments, the animals are placed in cold environments and
press panels in order to obtain heat. Young pigs exposed to cold soon
learned to press panel switches with their snouts in order to obtain radiant
heat from infra-red heaters suspended above their cages. When young pigs,
which had been trained to operate the heaters, were exposed at a range of
ambient temperatures from 10°C to 40°C it was found that the rate at which
they turned on the heaters declined markedly at 25°C (Baldwin & Ingram,1967).
The fact that they ceased to operate the radiant heaters at an ambient
temperature of 25°C is of physiological interest because at this temperature
young pigs about 2 or 3 months old enter their thermo-neutral zone, in which

the ambient temperature makes minimal metabolic demand upon the animal. There would be little physiological advantage in operating the heaters above an ambient temperature of 25°C and it can be assumed that the pigs were cool at 20°C but comfortable at 25°C.

Shorn sheep can also learn to press for heat and it has been shown that sheep can compensate very accurately for changes in the intensity of the radiant heat. The shorn sheep were trained to operate the infra-red heaters and were exposed for 48h at an ambient temperature of 10°C with either 900 watts or 1800 watts of infra-red heaters suspended above them. It was found that the sheep halved the duration of heating obtained when the intensity of radiant heat was doubled. This result illustrates the precision of the neural mechanism which controls thermoregulatory behaviour (Baldwin, 1975).

REFERENCES

Baldwin, B.A. 1975. The effect of intra-ruminal loading with cold water on thermoregulatory behaviour in sheep. J. Physiol., 249, 139-152.
Baldwin, B.A. 1979. Operant studies on the behaviour of pigs and sheep in relation to the physical environment. J. Anim. Sci., 49, 1125-1134.
Baldwin, B.A. 1981. Shape discrimination in sheep and calves. Anim. Behav., 29, 830-834.
Baldwin, B.A. and Ingram, D.L. 1967. Behavioural thermoregulation in pigs. Physiol. & Behav., 2, 15-21.
Baldwin, B.A. and Meese, G.B. 1977. Sensory reinforcement and illumination preference in the domesticated pig. Anim. Behav., 25, 497-507.
Baldwin, B.A. and Start, I.B. 1981. Sensory reinforcement and illumination preference in sheep and calves. Proc. R. Soc. Lond. B., 211, 513-526.
Dawkins, M.S. 1977. Do hens suffer in battery cages? Environmental preferences and welfare. Anim. Behav., 25, 1034-1046.
Dawkins, M.S. 1980. Animal Suffering: The Science of Animal Welfare. London. Chapman & Hall.
Duncan, I.J.H. 1978. The interpretation of preference tests on animal behaviour. Applied Anim. Ethol., 4, 197.
Hughes, B.O. 1976. Preference decisions of domestic hens for wire or litter floors. Applied Anim. Ethol., 2, 155-165.

DISCUSSION

Chairman: M. Zanforlin/Italy

The lively discussion that followed the contributions showed clearly the general interest and relevance of ethological indicators to assess the welfare of animals. Of particular interest were the observations and the questions concerning the technique of operative conditioning by which the animals can determine the environmental conditions that they "prefer". With respect to the interesting fact that the light seeking behaviour of calves is not connected to a circadian rhythm, it was suggested that the motivation inducing it might be indirect. Light may offer to the animal the possibility of satisfying the need for social stimuli and interactions.

A second topic on which the discussion centered was the possible interpretations of stereotypies. Both possibilities that the stereo-typies were due either to low level stimulation (i.e. boredom) or to a persistant high level of stressful stimuli, were challenged on the ground that the stereotypies are always very specific and simple motor-patterns. It was suggested that stereotypies could be merely "escape" behaviours. Thus, even persistant chewing behaviour observed in cattle and pigs could be interpreted as a type of "espape" behaviour.

Apart from the specific "motivation" behind the various stereotyped behaviours, everybody agreed that they are manifestations of a low welfare status.

SESSION II

SIGNIFICANCE OF INDICATORS RELEVANT TO ANIMAL WELFARE
Pathological indicators

Chairman: R. Moss

AN ECOPATHOLOGICAL APPROACH FOR ASSESSING RISK FACTORS
IN INTENSIVE REARING SYSTEMS :
EXAMPLES FROM PIG PRODUCTION

J. P. TILLON, D. V. M.

Ministère de l'Agriculture, Direction de la Qualité
Services Vétérinaires
Station de Pathologie Porcine, B. P. n° 9 - 22440 Ploufragan
France

ABSTRACT

A method based upon assessment of "risk factors" derived from ecopathological studies and used to prevent enzootic herd diseases is presented.
The author submits that method to the behavioural scientists because pathology and behaviour problems can be considered as two kinds of manifestation
of the disharmony of an animal intensive production system.

INTRODUCTION

During the past 10 years, reproductive performances in pigs have not
stopped progressing. In France, the average annual number of piglets by sow
has moved from 17 to 20 (in a sample of 170 units analysed from 1971 to
1981). Between these dates the value of pork has gone down slightly at
constant prices, while the cost of feed has risen. This means that the
productivity gains achieved by pig farmers were indispensable for the maintenance and for the development of pig-meat production, allowing consumption in France to rise from 27 to 45 kg per inhabitant annually.

Striving for higher productivity and a numerical increase in the pig
herd, pig farmers have adopted intensive rearing methods, among which we
may cite : early weaning, organisation of the sow herd, selection of animals having high growth potential... Economies have been achieved in the
cost of feeding (sow rationing), in the cost of housing (tethering sows,
high stocking ratios) and in the cost of labour (elimination of bedding,
introduction of slatted floors) and many units have automatic feed and
water distribution system.

The existence of modern pig units often very large, with between 50
and 200 sows is made possible by the success of the fight against serious
diseases such as Swine Fever. Other contagious diseases such as Aujeszky's
or viral gastroenteritis are widespread, thanks to the close proximity of
pig farms and the frequent movement of commercial animals : they flourish
from time to time, but do not constitute a permanent threat to animal health.

Far more serious from the economic stand point are <u>enzootic herd disea-</u><u>ses</u> which have become ever more common and which affect a certain number of pig units on a more or less permanent basis : for exemple coliform scours at weaning, enzootic pneumonia, atrophic rhinitis, scouring in the unweaned piglet, fetal mummification, MMA syndrome, urinary infections, small lit-ters, return to heat... There is virtually no unit where one or more of these problems does not occur repeatedly.

The economic losses (on average around 20 % of the total cost of pro-duction) occasioned by these health factors explain the need for the vete-rinary surgeon in pig production. <u>Antibiotics and antimicrobials</u> were for a long time the mainstay of therapeutic and prophylactic actions. Because of their poor efficacy, the build-up of resistance, the high cost of treat-ments, and because of consumer pressure, we have been studying other solu-tions, divising strategies based upon an <u>ecopathological approach</u> of the health problems.

After presenting the basic ideas of this approach, I intend to take as an illustration of its use the digestive troubles of weaner piglets. In a third section I shall show how this approach can be used to define the conditions necessary for animal welfare.

BASIC IDEAS IN THE ECOPATHOLOGICAL APPROACH TO HEALTH PROBLEMS

The ecopathological approach is intended to be used in the analysis of a multi-factor problem. A first difficulty is always to reply to the follo-wing question : <u>is it a multifactor affliction</u> ? In the example of diges-tive troubles in weaners, innumerable publications have demonstrated that coliform organisms can cause diarrhoea, oedema, and death. On the other hand no strain has ever been isolated which produces invariably and by it-self the problems which one finds in the field. No vaccine has been effec-tive against this pathological condition. So we conclude that the coliforms are influenced by others factors. We know that the digestive troubles are not caused by any other bacterium, because we have been totally unable to identify one regularly in SPF pigs contaminated by sick animals. Further-more, the trouble is often prevented simply by removing piglets from their original milieu : which proves that environmental factors play a role in observed digestive troubles at weaning.

It is unnecessary to detail the other examples, which prove that the majority of the herd diseases mentioned earlier are conditioned by numerous different factors. The infective agent which is potentially pathogenic is not capable of producing, alone and unaided, the observed disorders.

The ecopathological approach considers the animal unit as a whole, as a single entity. There is not one animal or one category of animals which is affected : it is rather the system which expresses a degree of disharmony between its constituent elements.

This statement is extremely important, for the result is that we fuse together all pathological signs, and we retain only this one single idea : the symptom which I observe warns me that something is going wrong in this (pig) production unit.

It is of minor importance whether the observed symptom is that of piglets coughing or scouring, whether sows are aborting or whether they have no milk... The ecopathological approach is global, and analyses the system as a whole : and this is where it differs from the usual clinical approach, which is based on the classification of individual symptoms.

If this method refuses to accord more importance to one clinical symptom rather than another, the ecopathological approach also refuses to consider one influence as more important than any other.

When a veterinarian arrives at a farm, he usually tries to identify a pathogenic organism using whatever means he may have at his disposal, while he ignores whole areas of the animals environment (housing and food especially).

The ecopathological approach requires the use of a rigourous and exhaustive observation matrix which may look rather like a questionnaire. The questionnaires which we use at the Pig Pathology Research Station at Ploufragan are constructed from an hexagonal model which distinguishes causative influences (or variables) and resulting variables.

The perfect pig herd can be represented by a regular hexagon enclosed within a circle, the radius of which has its length equal to unity. This circle is called "optimised causative variables circle" (Fig. n° 1).

Each of the six angles of the hexagon represents a causative variable and all of them are at the same distance from the circle center. The performances (technical, economic, sanitary) of this herd can be represented by a second circle enclosed within the hexagon. This circle is the largest regular geometrical form which can fit into the hexagon. This diagram represents the perfect, or ideal, pig unit.

FIGURE 1   Diagram of the perfect pig unit.

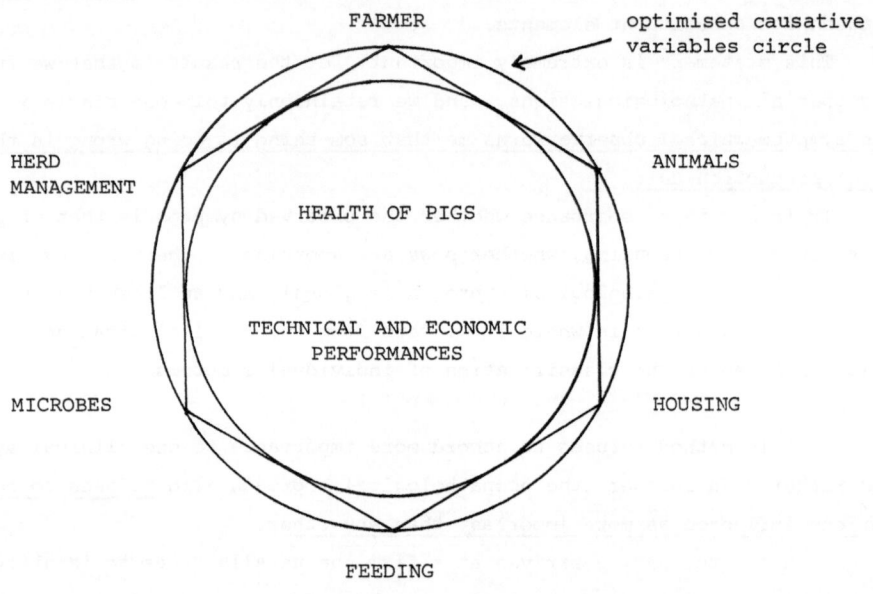

Frequently all the <u>causative variables are not optimised</u> : several may be
located somewhere along their axes, between the circumference and the center of the circle. In the figure 2, the irregular hexagon no larger fits
into our "optimised causative variables circle". The larger concentric circle enclosed within the hexagon is limited by the angle which is nearest
the center : and this is the principal limiting factor. So we can see that
the resulting variables represented by the surface of the enclosed circle
will not be very good. It is easy also to understand that no long-term
solution can be found without improving the two limiting variables [farmer]
and [microbial environment]. Having improved these two variables, it will
be also necessary to better the variables [feeding] and [herd management]
if we are to achieve the best possible results from the production unit.

FIGURE 2   Diagram of an ordinary pig unit.

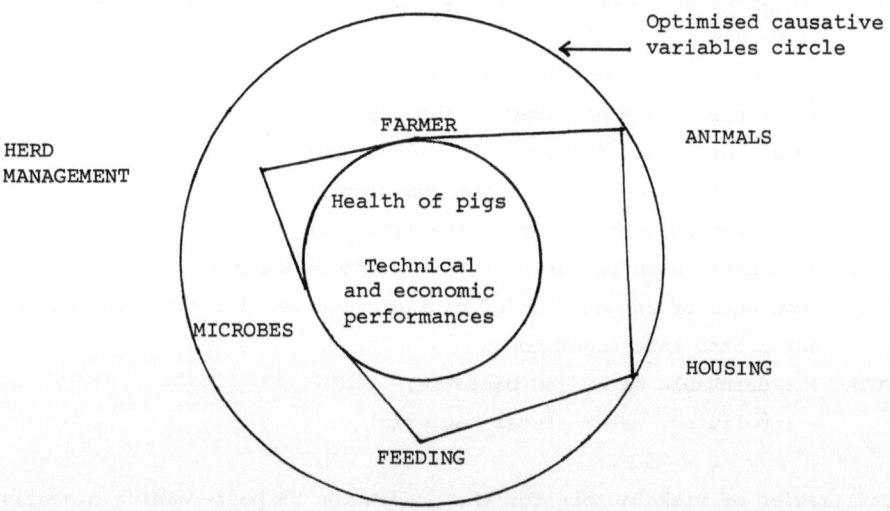

To be effective, the ecopathological approach needs <u>to compare the pig unit being considered point by point</u> with a sample group of herds which have already undergone the same process of examination. It is therefore essential to create <u>a network of sample herds</u> as a control. The observations obtained from this sample are <u>fed into a computer</u>, which is the only way possible to interpret the herd comparisons. In the Pig Pathology Research Station at Ploufragan we have three networks of sample herds participating in our epidemiological studies :

A <u>primary network</u> of 30 pig herds observed continuously. They are used as sentinels, telling us about the sanitary events which can need particular studies.

A <u>secondary network</u> of 150 herds which deals each year with a special study (in 1982 : reproductive disorders). For each herd, an observation team is composed of the farmer and his vet, both of them having to register the information contained in an adapted questionnaire. At the end of the observation period, all the information is transmitted to the Pig Pathology Research Station for interpretation (research of risk factors).

A <u>tertiary network</u> with a variable number of herds which apply the results

KCA   Level of daily feed (sows)

      (1 : low   2 : medium   3 : high)

NBJ △ Frequency of thermal amplitude higher than 6° C during the first
      21 days after weaning

      (1 : O   2 : one to four days   3 : more than 5 days)

PAIL   + : straw   - = no straw   at weaning period (piglets)

QKS   Amount of creep feed eaten by piglets before weaning
      (from 1 to 4 according to the quantity)

PMS   Mean weaning weight (increasing from 1 to 4)

DIAR   + néonatal scouring or white scour before weaning

GEC   Occurence of an outbreak of contagious viral diarrhoea in the herd
      during the last eight month

POTA   + : drinkable water (no bacteria)

      - : polluted water (fecal bacteria)

Utilization of risk factors for the prevention of post-weaning digestive troubles

Whenever a group of piglets is weaned, the probability of success is related to the number of satisfying risk factors. To avoid troubles, it is better to improve the deficient conditions than to try to prevent the proliferation of coliforms in the bowel of the animals. After measuring the value or risk factors in a group of piglets and using the map of risk factors, we can estimate the probability of success in weaning ; by eradicating the worst variables one after one and replacing them by improved variables, it is possible to simulate another situation (figure 5), which could be the situation of the considered herd after the improvement of the weaning conditions. If the herd reaches the safe area of the risk factor map, we may consider that the prevention of post-weaning digestive troubles would be achieved.

Further developments

Several enzootic herd diseases may be avoided in pig farms, by using the risk factors method previously described : respiratory diseases, MMA syndrome, scouring in the unweaned piglet, urinary tract infections... Many vets and a lot of pig-farm technicians and farmers are applying this method in the field. The use of drugs in pig production has decreased and the efficiency of this preventive method is always satisfying, even if all the risk factors cannot be improved in the short-run.

Research of risk factors. At the term of numerous statistical processes we may establish ten selected variables (= "risk factors") among all the information related to causative variables. These ten variables are generally satisfying among the herds without post-weaning digestive troubles and not satisfying among the others, the problems being more and more acute when positive variables are replaced by negative ones. In other terms, the more you neglect these variables and higher the risks are when weaning the piglets. Considering simultaneously the ten risk factors all the herds may be located on a "map of risk factors" (fig. 4) and their location is comparable with the repartition obtained in figure 3 : the same symbols tend to be localised in the same area of the map, the herds where the troubles are mild (★) being concentrated towards the center of the map.

FIGURE 4   Map of "risk factors" (letters)
            and location of the herds (symbols)

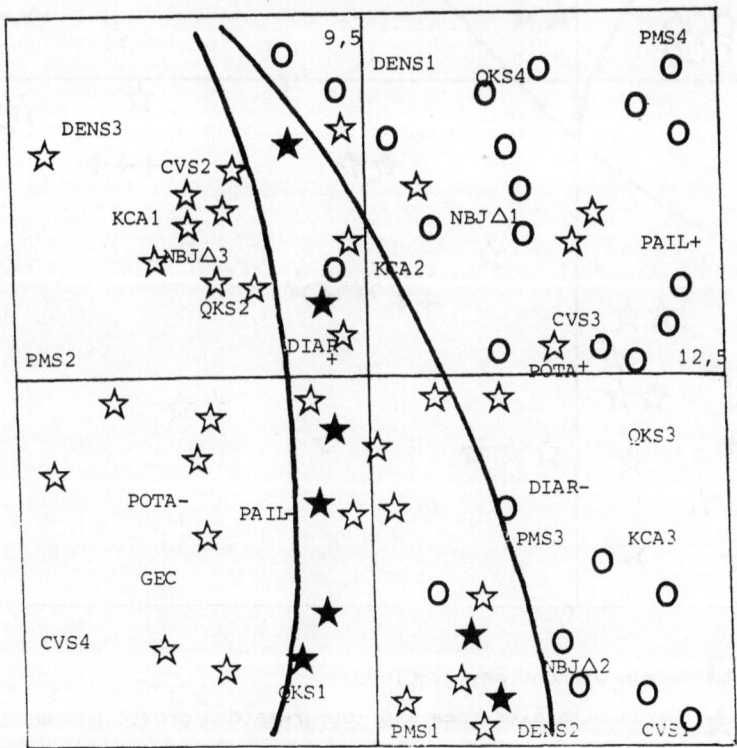

Legend :

DENS   Stocking rate of piglets after weaning

       (1 : low   2 : medium   3 : high)

132

. the herds with diarrhoea and (or) retarded growth problems, but no
mortality ★

. the herds where mortality is observed ; in most cases these herds
are also afflicted with severe diarrhoea and retarded growth ☆

FIGURE 3   Herd classification

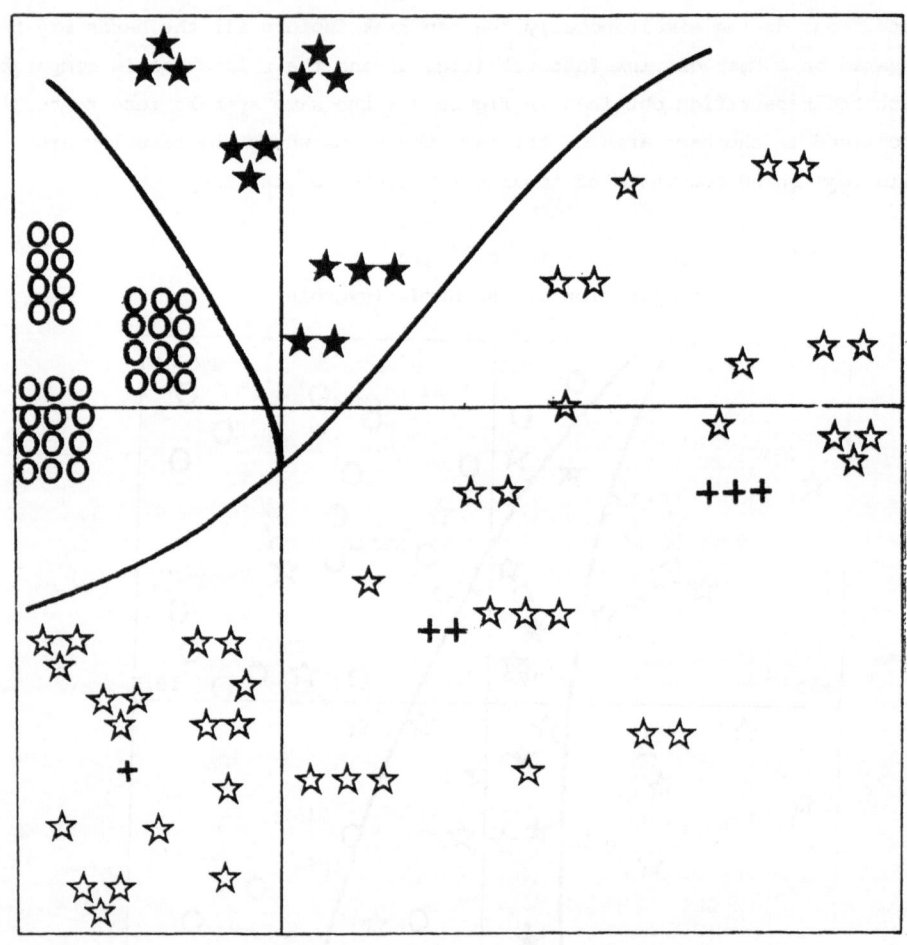

Legend

O   Herds without any problems

★   Herds with diarrhoea and (or) retarded growth but no
      mortality

☆   Herds with mortality (+, ++, +++ : severity of the post-
      weaning digestive troubles)

of the former studies to their own pathological situation (risk factors evaluation).

EXEMPLE OF APPLICATION : IDENTIFICATION OF "RISK FACTORS" IN THE POST-WEANING DIGESTIVE TROUBLES

As said previously, the post-weaning digestive troubles (including mortality, diarrhoea and retarded growth of piglets) are conditioned by numerous factors which appears as "risk factors". The problem may occur at any time if there are too many insufficiencies in the way the piglets are weaned. In order to prevent these troubles, the consulting vet has to find out what the insufficiencies are and correct them. But the main difficulty the vet deals with is to select among a lot of noticeable facts the "risk factors" which have to be eradicated first. To assess these risk factors it is necessary to carry out an ecopathological study in a network of sample herds.

Method

A hundred breeding pig herds constitute the sample of the study ; they are located in the same area and they may be compared in size and management. Half of them are regularly afflicted with post-weaning digestive troubles, whereas the other herds are free of this kind of disease. All these herds are subjected to the same investigation proceedings, a group of piglets (50 to 150 piglets) is observed from their birth on til three weeks after weaning and several daily records are made. At the end of the period 520 pieces of information have been recorded.

Data processing

The recorded data are fed into a computer and submitted to statistical methods, particularly the "principal components analysis". We may distinguish two steps in the data processing.

Herd classification. The retained criteria selected among resulting variables for measuring the importance of post-weaning digestive problems are : mortality, diarrhoea and retarded growth of piglets. Each herd is more or less affected by these troubles, all the combinations being possible between them.

In fact three main groups of herds are recognisable (Fig. 3) :
. the herds without any problems O

A wider utilisation of these methods of prevention based on the estimate of the risk factors depends on two conditions :

<u>Microcomputers</u> at the user's disposal enabling to locate a farm on a risk map and to carry out the evolution simulation,

A better training for vets and technicians both in <u>zootechnical knowledge</u> and in <u>methodology</u>. It is in fact rather difficult to give up the too famous analogy : disease = microb.

FIGURE 5   Simulation on a map of risk factors of the probability of occurence of post-weaning digestive troubles in a pig herd after improvement of the weaning conditions (1 to 5)

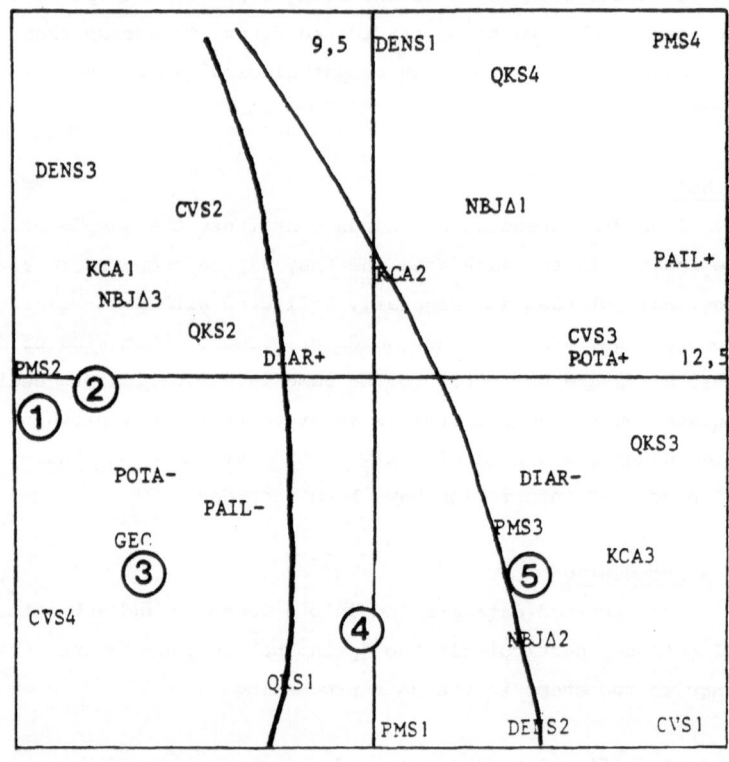

Legend :

① All the risk factors are deficient
② Only one of the risk factors is improved
③ Improvement of DENS (3 → 2), POTA (– → +) et KCA (1 → 2)
④ Improvement of other risk factors
⑤ The herd has reached the safe area

SUGGESTIONS FOR A MULTIFACTORIAL APPROACH OF THE WELFARE OF THE PRODUCTION
ANIMALS. EXAMPLE : TEETH-GRINDING OF THE SOW

The practice of health problems in animals lead us to consider the
symptoms, i.e visible or audible manifestations in the animals. During the
visit some behaviour (or so-called) problems can be observed. The ecopathological approach then may lead us to analyse them in relation to all
the environmental factors. Our experience in that area is still limited
since we have dealt almost exclusively with health problems. Nevertheless
I would like to put forward a few elements that might require the consideration of ethologists and make the dialogue with pathologists easier.

As an example I will consider the teeth-grinding phenomenom of the
sow from an ecopathological point of view.

The different elements of the problem

Teeth-grinding is a manifestation which draws the attention of the observer. It can be considered as a "message" betraying pain, a behaviour deviation,
or the unsatisfaction of the basic biological needs. This manifestation reveals a disharmony between the animal and its rearing environment. We can also admit that it is the consequence of unfavourable influence as a whole.

Teeth-grinding in sows may induce two types of reasoning :

A means (or different means) may be studied to get rid of this apparently pathological manifestation (freeing the sows from their ties, giving
them straw, or open air rearing),

Teeth-grinding may be a pretence to investigate the disharmonious elements of the ecosystem and try to improve them for a better comfort of the animal. In that case we will try to quantify the teeth-grinding (mumber, duration)
so as to reveal the influences which worsen or improve the situation. A comparison between sampled farms will thus reveal the "risk factors" for teeth-grinding.

An attempt to formulate an observation protocol for the ecopathological
analysis of the teeth-grinding of sows

Basis of comparison of pig-farms. In each sampled farm the teeth-grinding is
numbered before feeds and one hour after. The affected animals are identified;
other data about the beginning and the duration of the fits of teeth-grinding,
their occurence in time, in the presence of man, the association with other behaviour manifestations (bites on bars, pruritus) may also be noted down).

Formulation of an analytic protocol appliable to a herd of sows

The farmer :
- the quality of the caretaking,
- the relationship between the man and the animal,
- the time spent among the animals.

The herd :
- the size of the farm,
- the breed, the origins,
- the demographical structure of the herd,
- the condition of the sows,
- the motivity of the sows (number of animals lying or standing at different moments of the day),
- the frequency of urination,
- the quality of faeces (constipation),
- urine analysis (infection),
- limb observation (locomotor problems),
- the level of performances.

The buildings :
- temperature in the building,
- frequency of important differences of temperature,
- appraisal of the quality of the ventilation,
- type of flooring. Straw litter,
- restraining apparatus,
- noise intensity,
- size of the sty, the orientation,
- number of animals penned together,
- the light.

The feed :
- quantity of feed distributed,
- means of feeding,
- frequency of the feeds,
- alimentary formulation,
- fineness of the grinding
- mineral and vitamins additives,
- average speed of ingestion of the feed (sows),
- quality of the drinking water,
- quantity of water,
- watering system.

Microbial infection and health condition :
- scab,
- internal parasites,
- abcesses, chronical lesions,
- respiratory or digestive pathology,
- search for antibodies,
- investigation of culling sows at the slauthter-house : lungs, stomach, kidneys, bladder, joints, udder,
- reproduction and farrowing pathology.

Rearing management :
- rearing methods of the young sows,
- age at first covering,
- weight at first covering,
- hygiene of the sties,
- preparation to the farrowing,
- first cares to the piglets,
- vaccination schedule,
- vermifugation.

This inventory, which is not exhaustive, may be the basis for a discussion and we must say to its merit that it makes compulsory for the observer a global approach of the problem, for from my own observations, each of the afore-cited elements can be linked to the observed problem.

Bringing forward the obviousness of the risk factors. Processing the data from an ecopathological investigation of teeth-grinding of the sows would certainly bring forward the fact that some circumstances are regularly associated to this trouble ; there are indeed pig-farms where it never occurs while others are permanently affected.

If it appears for instance, that teeth-grinding is associated to :
- the restraining apparatus of the sows,
- thermic discomfort (cold sties and frequent changes of temperature),
- chronical vesical problems,
- constipation,
- high frequency of locomotor troubles,
- too early covering of the sows,
- the existence of auricular scab.

We may suggest prophylactic rules which would not only solve the pro-

blem of teeth-grinding, but also improve the state of health and the level
of performances. To restrict the problem to the sole aspect of restraining
apparatus is far from satisfying, even if the latter is itself linked to
the other variables (urinary infections, constipation, scab...). In this
case, it is not only the restraining apparatus that must be revised, but
all the elements of the rearing management which must be coherent with this
method of restraining.

CONCLUSION : THE RELATIONSHIP BETWEEN PATHOLOGY AND ANIMAL BEHAVIOUR

Whether we are behavioural scientists or pathologists, we cannot but
be surprised at the way in which production animals are sometimes treated
to day.

The option which is adopted of intensive production, with high animal
concentrations puts in question all the standards with which we ourselves
were brought up. At the same time, the search for greater productivity appears to diminish the traditionnal high quality of our daily foods.

What attitude should we adopt ? It seems to us that the repressive
attitude which have always ruled our Veterinary Services are not well adopted to the context of keen commercial competitiveness, of continued economic activity and of the required evolution in the living standards of the
producers. On the other hand, a constructive attitude, based upon sensible
professionnal advice -which corresponds to the real context of our daily
life, and especially to our economic contraints- : such an attitude will be
accepted by producers, and it will lead in due time to a change in their
common practice (for example : a reduction in anbitiotic use, and a real
improvement in the welfare of our production animals). In this way, the
ecopathological approach and the derived use of risk factors methods allow
to help the farmers to solve the problems which occurs in intensive pig
production in adapting their means to their aims.

COMPARATIVE MORTALITY AND MORBIDITY RATES FOR CATTLE ON SLATTED FLOORS AND IN STRAW YARDS.

J. Hannan and P. A. Murphy

Department of Veterinary Preventive Medicine and Food Hygiene
University College Dublin 4.

ABSTRACT

The morbidity rate for fattening cattle in slatted floor houses was twice as great as that for their counterparts in traditional straw yards. Eye and skin diseases were more prevalent in slatted floor houses while clinical parasitism was more frequently recorded in cattle in straw yards. Lameness was the most frequently recorded disease in both husbandry systems. This disease was much more prevalent and more severe in cattle on slats. The number of lame animals increased when trough space was limited and when the non-slatted area in slatted houses was extensive. Morbidity and mortality rate may be useful in monitoring the welfare of animals in different husbandry systems.

INTRODUCTION

The proportion of two-year old beef cattle kept in houses with slatted floors during the period November to March increased from less than one per cent in 1969 to IO per cent in I977. Because of this change it was decided to monitor the morbidity and mortality rates in nine of these units and in nine conventional straw yard units over a two year period. The cattle were mainly Friesians (50 per cent) and Herefords (40 per cent) bullocks aged approximately two years of age and fed on silage ad libitum and some grain. The information presented in this paper was collected by personal interview with the aid of a standard questionnaire at three weekly intervals and disease records were compiled from reports of the diagnoses made by veterinary surgeons or from personal observation by one of the authors (Murphy, I979).

RESULTS

Losses resulting from deaths from natural causes, euthanasia which was performed on humane grounds and from salvage of animals that were sent to abattoirs are presented in Table I.

TABLE I   Deaths in Housed Beef Bullocks Expressed as a Percentage of Cattle at Risk.

	Slatted Floors	Straw Yards
Cattle at Risk	I2OIO	2882
Source of Loss		
Mortality	0.I2	0.2I
Euthanasia	0.OI	0.03
Salvage	0.42	0.24
TOTAL DEATHS	0.55	0.48

140

The overall mortality in both systems was about one animal in 200: the
mortality rate was higher from natural causes in straw yards, whereas
more cattle were sent for salvage from slatted floor houses.

The incidence of clinical disease in bullocks in slatted houses and
on straw yards is presented in Table II.

TABLE II   Disease Incidence in Bullocks kept on Slats and on Straw
expressed as a Percentage of Cattle at Risk.

	Slatted Floors	Straw Yards	Significance of Difference
Cattle at Risk	I2OIO	2882	
Disease Cateogry			
Lameness	4.75	2.43	P < O.OOI
Eye Disease	2.09	0.97	P < O.OOI
Skin Disease	O.9I	0.07	P < O.OOI
Acute Ruminal Impaction	0.38	O.I7	n.s.
Injury	0.30	0.24	n.s.
Abscessation	0.27	O.IO	n.s.
Clinical Parasitism	0.25	0.80	-P < O.OOI
Enteritis	0.23	0.00	P < O.O5
Respiratory Disease	O.I6	O.IO	n.s.
Other Diseases	0.39	0.54	n.s.
ALL DISEASES	9.73	5.42	P < O.OOI

While lameness was the most common disease observed in both husbandry
systems the incidence was twice as great on slatted floors.   Eye and
skin diseases were also significantly more prevalent in straw yards.
The overall incidence of disease was almost twice as great in slatted
floor houses.

Lameness cases were classified into four categories viz. mild,
moderate, severe or extensive depending on the length of time that
elapsed before the affected animal was considered fit to return to their
pens of origin.

TABLE III   Severity Ratings of Lameness Cases expressed as a
Percentage of Lameness Cases.

	Slatted Floors	Straw Yards	Significance of Difference
Number of Cases of Lameness	788	75	
Category			
Mild	7.9	I3.7	-P < O.OI
Moderate	45.4	78.3	n.s.
Severe	39.5	8.0	P < O.OOI
Extensive	7.2	0.0	P < O.O5

Not only was lameness a more frequent occurrence on slatted floors but cases were also more severe or extensive (47% compared to 8% on straw yards).

The distribution of lesions associated with clinical lameness is shown in Table IV.

TABLE IV   Lesions associated with Lameness expressed as a Percentage of all Lameness Cases.

Lesion	Slatted Floors	Straw Yards	Significance of Difference
Total cases of Lameness	788	75	
Septic Traumatic Pododermatitis	42.6	1.3	P < 0.001
Cellulitis of the Limb	21.5	0.0	P < 0.001
Aseptic Traumatic Pododermatitis	15.2	34.7	P < 0.001
Necrotic Lesions	9.8	45.3	-P < 0.001
Diffuse Aseptic Pododermatitis	5.5	13.3	-P < 0.01
Tendon and Muscle Injury	2.4	5.4	n.s.
Arthritis	1.5	0.0	n.s.
Osteitis and Fractures	1.5	0.0	n.s.

The most frequently observed lesion in cattle on slatted floors was septic traumatic pododermatitis which was the result of hoof penetration with infected material.   This caused painful abscess formation within the hoof.   The second major lesion associated with slats was cellulitis of the limb while necrotic lesions and aseptic traumatic pododermatitis were more commonly encountered in cattle in straw yards.

Among the factors which appeared to predispose to lameness in cattle on slatted floors was the trough space available for each animal (Table V).

TABLE V   Incidence of Lameness as related to available trough space.

Available trough space mm/head	Number of units	Number of cattle at risk	Percentage incidence of lameness	Significance of difference
300 - 450	4	4800	5.4	
451 - 600	3	1978	3.34	P < 0.001
601 - 750	2	540	3.15	

A second factor that was associated with the incidence of lameness was the width of the non-slatted area of the pens in slatted houses.

TABLE VI   Incidence of Lameness related to the Non-Slatted area in Pens.

Width of Non-slatted area mm	Number of units	Number of cattle at risk	Percentage incidence of lameness	Significance of difference
0 - 300	3	1958	2.25	
				n.s.
301 - 900	3	1489	3.22	
				P< 0.001
901 - 1850	3	3871	6.46	

When the non-slatted area in the pens was greater than 900 mm wide there was a significant increase in the incidence of lameness.

CONCLUSIONS

Lameness is the most common clinical disease in fattening cattle in Ireland and in terms of incidence and severity it is a serious problem in cattle housed on slatted floors.   It's prevalence can be reduced by proper design of units so as to provide adequate trough space and wall to wall slatted floors.   From the data presented it is suggested that mortality and morbidity rates be used to evaluate newly introduced systems of husbandry and to determine their effects on animal welfare.

REFERENCES

Murphy, P. A. 1979.   Diseases of beef cattle housed intensively with particular reference to lameness.   PhD Thesis, University of Dublin.

Institut für Veterinär-Pathologie der Freien Universität
Berlin

PATHOLOGICAL APPROACHES TO EVALUATE CALF BOXES UNDER

ANIMAL WELFARE ASPECTS

by

K. DÄMMRICH [+]

I. CLAW LESIONS

The hind and front claws of calves were examinated macroscopically and microscopically. The macroscopical examination
includes the size of claw sole. The histological examination
of the claw sole includes horne thickness, structure and
the incidence of lesions.

Structure of sole horne

The corium of the sole shows a high papillary structure.The
epidermis is arranged in horny channels. Each channel
contains the top of a corial papilla and is surrounded by
interpapillary horne. The horne is developed by the epidermis. The proliferating part of the epidermis is the
stratum basale. The stratum basale produces cells, which
forme the horne by means of differentiation. Stratum
spinosum and stratum lucidum are differentiated cell layers.
Later on the cells of stratum lucidum become necrotic. The
necrotic cells are the superficial stratum corneum. The
quality and strength of horne depends from the development
of desmosomes in the stratum spinosum and from the development of eleidine in the stratum lucidum and from the
arrangement of the tonofibrills in the stratum corneum.

Lesions of sole horne

Circulatory disturbances were the main lesions in the sole

[+] In cooperation with Prof. Dr. J. UNSHELM and
Dr. U. ANDREAE: Institut f. Tierzucht und Tierverhalten
der Bundesforschungsanstalt für Landwirtschaft, Marien-
see/Trenthorst.

of claws. They were represented by acut focal hemorrhages, which occured in the corial papillae. The blood filled the proximal part of the horny channels. The content of the channels, the corial papilla, became necrotic. Together with the hemorrhages the structur of stratum basale, stratum spinosum and stratum lucidum was disturbed. Regressive changes, like focal necrosis of basal cells and inter- and intracellular edemas in the stratum spinosum et lucidum, were observed. The decrease of keratinisation was characterized by the loss of eleidine granules and by the loss of basophilia: dyskeratosis.

With the growth of horne to the surface of the sole the hemorrhages lost the contact to the regenerated corial papillae. Now the part of the horny channels, which contains the blood, lies in the stratum corneum. With blood filled horny channels lay together in stripes, which were sur- rounded by dyskeratotic horne. Sometimes two or three stripes of blood filled channels occured in the sole horne as a sign of repeated hemorrhages. Such older hemorrhages were called chronic hemorrhages.

Single boxes with wooden slatted floor compared to single boxes with concrete floor and straw

We examined the claws of the right forelimb and of the right hind limb of male calves in the age of 26 weeks. 13 calves were kept in single boxes with wooden slatted floor and 14 calves were kept in single boxes with concrete floor and straw.

The same type of lesions was observed in the two groups, hemorrhages together with dyskeratosis. The frequency of lesions was different between the two groups. In calves, which were kept on slatted floors, only in 23 % of the examinated claws lesions occured. But in calves, kept on concrete floor, 71 % of the examinated claws were altered. The frequency of lesions was higher in the hind limbs.

Hemorrhages and dyskeratosis are the result of traumatic injuries. The comparative investigation of the sole horne resulted in a varying structure of the sole horne in the two groups. Calves on slatted floor showed a regular arrangement of dense uniforme bundles of tonofibrills. The narrow horny channels occured with an amorphous basophilic shadowy structure. In calves on straw the tonofibrills were arranged irregulary; they formed fragmentary bundles with wide interspaces. The horny channels were wide and they contained the basophilic medulla with persistent nuclei. The structure of horne depends from the water content. Low water content corresponds with dense bundles of tonofibrills and hard consistency of horne. High water content corresponds with incompact bundles of tonofibrills and softer consistency of horne.

The water content of horne determines the abrasion of the sole horne. Low water content of horne on slatted floor led to a higher degree of abrasion with the result of thinner sole horne and the increased size of the sole. Otherwise in calves on straw the horne contained a higher content of water. The more rubberlike consistency prevents the horne from abrasion. The result was the thicker horne and the smaller size of the soles.

The protection against traumatic injuries of the epidermis depends from the water content of the horne. In calves on slatted floor the hardness of sole horne with low water content protects the corial membran much better against traumatic injuries than the more soft horne with high water content in calves, which were kept on concrete floors with straw.

2. Single keeping compared to keeping in groups of 4 calves

The examinated calves were kept in boxes with wooden slatted floors. The age of the calves ranged between 16 and 26 weeks.

Differences in size, shape and structure of the sole horne did not exist. The findings were very similiar to the results in calves, which were kept on wooden slatted floors in single boxes.

Chronic hemorrhages together with dyskeratosis occured. The histological appearance of the lesions was described previously. The hind legs showed more regressive changes than the forelegs.

In single kept calves the frequency of claw lesions was lower than in calves, which were kept in groups. The difference is not caused by different structures of the sole horne in the two groups. The difference can be explained with the higher locomotory activity of calves in grouping condition. The higher locomotory activity leads more often to collisions and false steps on the slatted floor with contusion of the claw epidermis.

Finally we can summarize the results of our investigations as followed. We compared calves in single boxes with wooden slatted floors to calves in single boxes with concrete floor and straw. The incidence of claw lesions was lower on slatted floors. It depends from the decreased water content of claw horne together with increased mechanical resistence against traumatic injuries. Otherwise we compared calves on wooden slatted floors, which were kept in single boxes or in groups. The frequency of claw lesions was higher in group keeping than in single keeping. The higher incidence of claw lesions was caused by the increased locomotory activity of the calves in group keeping.

References

Brentano,G., K. Dämmrich u. J. Unshelm, 1979: Untersuchungen über Gelenk- und Klauenveränderungen bei auf Lattenrosten und auf Stroheinstreu gehaltenen Mastkälbern.
Berl. Münch. Tierärztl. Wschr. 92. 229 - 233.

K. Dämmrich, J. Unshelm, U. Andreae u. R. Bader, 1982: Untersuchungen über Klauenveränderungen bei in Einzeltier- und Gruppenhaltung aufgezogenen Mastkälbern.
Berl. Münch. Tierärztl. Wschr. 95. 21 - 26.

Table 1: Size and structure of claw soles

sole	slatted floor		floor with straw	
	foreleg	hind leg	foreleg	hind leg
size (mm^2)	2145	1707	1454	1344
thickness (μm)	4737	4899	9670	9718
structure horny channels	narrow;amorphous basophilic shadowy structure		wide; basophilic medulla with persistent nuclei and interpapillary lamellae	
tonofibrills	regular arrangement of dense uniform bundles		irregular arrangement of fragmentary bundles in varying size and shape	

Tables 2 and 3: Claw lesions

SINGLE BOXES WITH WOODEN SLATTED FLOOR
COMPARED TO
SINGLE BOXES WITH CONCRETE FLOOR AND STRAW

lesions of claws (% of examinated claws)	slatted floor		floor with straw	
	foreleg	hind leg	foreleg	hind leg
acute haemorrhages	--	3,6	3,5	3,5
chronic haemorrhages	--	11,5	14,0	21,5
dyskeratosis	7,2	19,2	50,0	46,4
total of alterated claws	23,1		71,4	

SINGLE KEEPING COMPARED TO KEEPING IN GROUPS

lesions of claws (% of examinated claws)	single keeping		keeping in groups	
	foreleg	hind leg	foreleg	hind leg
chronic haemorrhages	10,2	28,3	45,9	53,1
dyskeratosis	5,0	10,0	--	10,2
total of alterated claws	27,5		54,6	

II. PYLORIC ULCERS

The occurance of pyloric ulcers in calves is very high. Many reports exist. Pyloric ulcers occur in young calves. Later on in young cattle the pyloric ulcers heal up. The pathogenesis of the ulcers is unknown, because the pathological investigations were done only macroscopically and only in chronic stages of ulceration. Speculations about the etiology of pyloric ulcers discuss such causes as local circulatory disturbances, incarceration of straw and hay bundles in the pyloric channel, mechanical irritation of the pyloric mucosa by roughage, hyperacidic reaction of the abomasal content, decreased secretion of mucus and nervous influences in stress.

We investigated macroscopically and microscopically the pyloric region. The age of calves ranged from 16 to 26 weeks. The calves were kept in single boxes or in groups. The calves were fed mainly with milk substitute. The calves of the last group were suckled by cows. The first group did not receive roughage. In the other groups roughage was added. Roughage was ground straw pellets or chopped straw. The calves of the last group took up food of the mother cows.

At first we investigated the size of abomasum and the histological structure of the abomasal wall. The result of this investigations: the size and the structure of abomasum did not depend on the different feeding systems, but size and structure depend on the age of the calves.

Lesions of the pyloric mucosa

Primary the lesions were localized on the pyloric torus or secondary in the pyloric channel, which are surrounded by the sphincteric muscle.

Lesions occured in all of the examined groups. The calves were affected with an incidence of 72 till 1oo %. Significant differences between the kind of feeding and the occurance of lesions did not exist. The same lesions were found in calves, which had no access to roughage.

Morphology of pyloric lesions

Secretion of mucus was disturbed. The reduced number of goblet
cells in pyloric glands led to the decreased secretion of
mucus, in consequence the pyloric mucosa was not protected
enough. Increased secretion of mucus was mainly a secondary
protective reaction. It occured in the surrounding area of
inflammation, ulceration or hemorrhages.

Circulatory disturbances occured very often. Local hyperemia
and dilatated capillaries were the first signs of circulatory
failures. Later on the stagnation of the blood flow in
dilatated capillaries lead to the formation of hyaline
thrombi. Thrombotic capillaries were situated in the top of
the mucosal villi. Dilatation and thrombosis of capillaries
were accompanied by edema and hemorrhages.

Erosion of the mucosa started with localized acute necrosis
of the villi. The necrotic villi are demarcated by inflamma-
tory cells. Repeated necrosis and autodigestion enlarged
small necrotic foci to extended erosions of the pyloric
mucosa.
If the necrotizing proceeding is finished the mucosa is able
to regenerate. Only large and deep erosions, which include
the whole mucosa, heal up with scars.

Pyloric ulcers include mucosa and submucosa. Ulcers develop
from acute deep erosions. Granulation tissue demarcates the
ground of ulcers in acute and subacute stages. Very often
ulcers are progressive, because repeated necrosis – caused
by autodigestion – happens. Perforated ulcers did not occur
in the investigated calves.

Foreign bodies are vegetable fibres and fragments of calve
hairs, which intrude into the mucosa and cause surrounding
inflammation. Secondary the granulation tissue of ulcers
contains such particles. Sometimes the abomasal content
includes long blades or haulms. They can form bundles or more
globular phytobezoars.

Pathogenesis of pyloric ulcers

The development of pyloric ulcers can be deduced from the
physiological data about the abomasal or pyloric function
and from the microscopical findings.

The pyloric sphincter of young ruminants is more powerfull
than in older animals. After uptake of milk or substitutes
the sphincter is activated, if fluid passages into the
duodenum. The contraction of sphincteric muscle closes the
pyloric channel. The closure continues till the digestion
is finished. Repeated closure depends from the passage of
whey into the duodenum. The coagulated part of milk remains
in the abomasum over a long time, in calves 13 till 17 hours
were reported.

In veal calves large volums of milk substitute or milk lead
to increased duration and strength of pyloric constrictions
during abomasal digestion. The sphincteric muscle compresses
the pyloric mucosa, especially the mucosa of the torus. In
consequence of compression the mucosa becomes anemic. The
effect of compression can be increased, if bundles of haulms
or phytobezoars are incarcerated in the pyloric channel.

Compression of the mucosa with anemia leads to hypoxia in the
tissues. Over longer time ischemia causes hypoxic lesions in
the mucosa.

After complete digestion the pyloric sphincter becomes slack
and the abomasum is evacuated. The pyloric mucosa is supplied
again with blood. But hypoxic lesions persist and they develop
circulatory disturbances. Signs of hypoxic circulatory disturbances are dilatated capillaries with stagnation of blood
flow and formation of hyaline microthrombi. The capillary
walls become permeable. Hemorrhages and edema are the result.
The circulatory disturbances are localized in the villi. They
make worse the primary hypoxic lesions and hypoxic necrosis
of villi occurs. Autodigestion develops the focal necrosis
of villi to erosions and ulcers.

Conclusion: The pyloric ulcers develop from hypoxic lesions,
which are caused by abnormal long acting contracctions of the
sphincteric muscle, due to large volums of milk or milk
substitutes. Abomasal incarcerated phytobezoars intensify
the ischemia of the pyloric mucosa. Intruded foreign bodies
are only in few cases the cause of pyloric lesions.

Reference

Degen, B., 1982: Pathologisch-anatomische und -histologische
                   Untersuchungen zur Pathogenese der
                   Pylorusgeschwüre bei Mastkälbern.
                 Vet.med. Diss. FU-Berlin.

Table 4:

FEEDING OF CALVES	
1. milk substitute two times daily ad lib.	
2. milk substitute two times daily ad lib.	+ straw (floor)
3. milk substitute two times daily ad lib.	+ pellets of ground straw
4. milk substitute two times daily ad lib.	+ chopped straw
5. milk substitute 5oo ml portions ad lib.	+ chopped straw
6. milk ad lib. one or two calves were suckled by one cow	+ straw and hay

Table 5:

```
┌─────────────────────────────────────────────────────────────────┐
│ PATHOGENESIS OF PYLORIC ULCERS                                    │
├─────────────────────────────────────────────────────────────────┤
│                                                                   │
│  large volume of milk substitute ( 2o l/d )                       │
│  increased duration of abomasal digestion                         │
│  increased duration and strength of pyloric constriction          │
│  compression of the pyloric mucosa                                │
│          ↓ ←——— incarceration of phytobezoars                     │
│  local ischemia                                                   │
│  hypoxia                                                          │
│          |——— complete digestion – pyloric dilatation –           │
│          ↓    abomasal evacuation                                 │
│  dilatation of capillaries                                        │
│  hemostasis                                                       │
│  hyaline microthrombi – hemorrhages – exsudate                    │
│                        ↓                                          │
│              hypoxic necrosis of villi                            │
│                        ↓                                          │
│                     erosion                                       │
│                        ↓                                          │
│                      ulcer                                        │
│                                                                   │
└─────────────────────────────────────────────────────────────────┘
```

RESULTS OF A METHODICAL APPROACH WITH REGARD TO EXTERNAL LESIONS OF SOWS AS AN INDICATOR OF ANIMAL WELL-BEING.

R. de Koning

Research Institute for Animal Husbandry "Schoonoord"

P.O. Box 501, 3700 AM Zeist, The Netherlands

ABSTRACT

Described is the importance of the integument in relation to well-being. Upto now lesions of the integument are primarely considered as deviations from the optimal health state. The importance of the lesions in relation to well being depends upon the pathological severity of the lesions.

In this paper is shown that the environment, the behaviour and the health state of the animal have an influence on the integument. This influence can be direct or through each other. It is discussed that while behaviour and health do reflect in the state of the integument, lesions of the integument are indicators of behaviour and health. As behaviour and health are prime indicators of well-being, the state of the integument is a derived indicator for well-being. Lesions of the integument still are pathological deviations of optimal health. Therefore it is concluded that the state of the integument is a multidisciplinary indicator related to well-being.

INTRODUCTION

One of the most cited, and apparently accepted definitions of well-being is formulated by Lorz (1973). In his comment on the German Animal Protection Act (Tierschutzgesetz) he defines well-being as a state of physical and psychical harmony of the animal within itself and with its environment. This harmony is difficult to measure, and therefore is well-being as such hard to assess. However, Lorz continues to say that indicators of well-being are good health and normal behaviour. Deviations in health and behaviour can be detected by a veterinary - or an ethological approach. Therefore these two disciplines are the two prime auxiliary sciences for the determination of well-being. According to Lorz good health and normal behaviour presuppose an undisturbed course of life's processes. Deviations in the course of these processes are indications of deviations in health or behaviour, and can therefore be considered as secondary indicators for well-being. Physiological data, including production figures, are such indicators.

Lesions of the integument are deviations from the optimal health state of the integument and therefore from the optimal health state of the indi-

vidual, and thus form an indication for decreased well-being of that individual. The integument is only a small part of the body, and the lesions
are most often of little pathological importance. Therefore uptill now
lesions of the integument were considered to be an indication of well-
being of minor importance.

These lesions however, do not occur spontaneously. Both the environment, and the physical and psychical state of the individual do influence
the occurrence of lesions. Therefore the state of the integument might well
be of more importance in respect to animal welfare matters then only because of its pathological significance. The purpose of this paper is to
illustrate this relation between the occurrence of lesions of the integument and the environment, the animals behaviour and its health status, and
its potential value for assessing well-being.

The results on which this paper is based are obtained from a survey
on lesions of the integument of dry sows. These sows were kept in different systems of individual housing (crates; or tethered by neck collar,
neck harness, or shoulder girth). Approximately 5000 sows were systemati-
caly inspected for lesions of the integument according to an inspection
routine. This inspection routine is part of the Ekesbo Method, which is
intended to distinguish between different environments in relation to well-
being, and is based on lesions of the integument.

THE INFLUENCE OF BEHAVIOUR, HEALTH AND ENVIRONMENT ON THE STATE OF THE
INTEGUMENT

The state of the integument of these sows was influenced by three
main factors: the environment; the behaviour of the sows; and the state
of health of the sows.

1. Behaviour

Behaviour can affect the state of the integument directly (grooming
automutilation) or indirectly. While behaving, the animal is in contact
with its environment. This causes lesions. The nature of such lesions depends on the intensity of the behaviour, and on the point of contact of
the integument and of the environment. Furthermore behaviour can change
the environment and influence the health state of the animal. Figure 1
shows how the behaviour influences the occurrence of lesions on the neck
of 80 sows which were tethered by a neck collar for the first time.

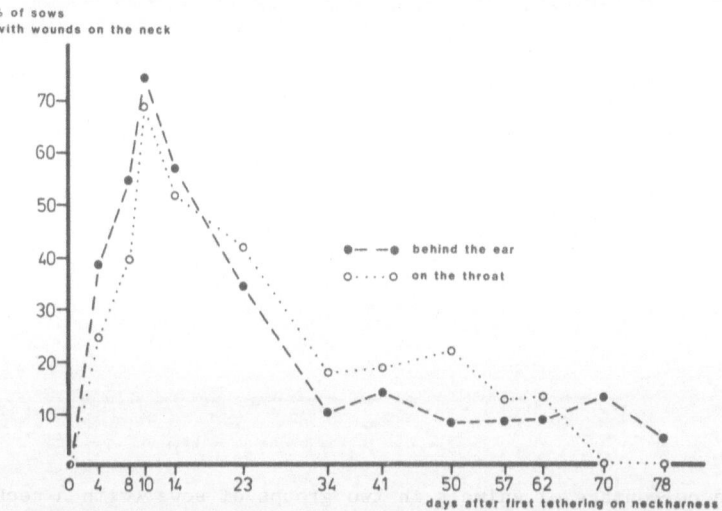

Fig. 1 The percentage of sows with wounds on the neck (behind the ear, and on the throat) when tethered by neck harness, at different days after first tethering.

Immediately after tethering a high increase of the occurrence of wounds takes place due to the resistance of the sows against tethering. After ten days the percentages drop because the animals get used to the system and the wounds heal. This is a clear example of how behaviour influences the frequency of lesions of the integument. In figure 2 results concerning wounds on the throat of the same group of sows are presented. Here, the group is devided into a group of sows with a neck circumference of more than 94 centimeters and a group with a neck circumference of less than 94 centimeters.

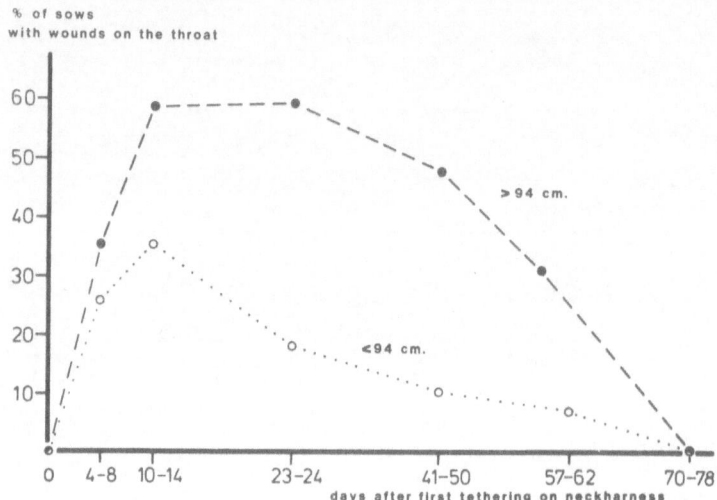

Fig. 2 The percentage of animals in two groups of sows (with a neck circumference of over 94 cm ●---●, and under 94 cm o···o) with wounds on the
throat when tethered by neck harness, at different days after first tethering.

In both groups immediately a high increase in the percentage of wounds
takes place, but in the group with a relatively wide neck harness (neck
circumference < 94 cm) the maximum level is only approximately half that
of the group with a relatively narrow harness, and the wounds do heal earlier in the first group. This makes clear that the environment has an effect on the level of lesions of the integument through behaviour.

2. Environment

As this latter example showed, environment and behaviour can be in-
terrelated while causing lesions. The environment can also have a direct
mechanical influence on behaviour, through which changes in the state of
the integument occur. An example of this route of influence is given in
figure 3.

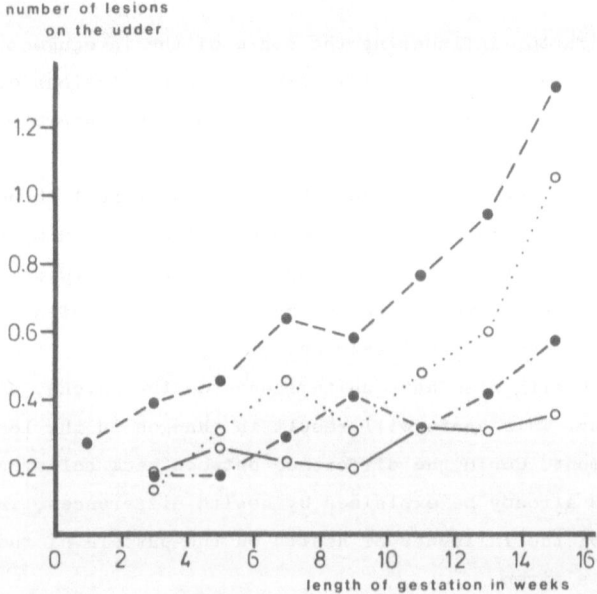

Fig. 3 The number of lesions on the udders of sows, tethered by shoulder girth (●---●), neck collar (o···o), and neck harness (●-·-·●), and housed in crates (o——o) during gestation.

The number of lesions on the udder of sows tethered by shoulder girth is significantly higher than that of sows tethered by neck harness or kept in crates. The number of lesions on the udders of sows tethered by neck collar is significantly higher than in sows tethered by neck harness. The difference between shoulder girths and crates or neck harnesses is most likely due to a mechanical hampering when getting up of sows tethered by shoulder girth. The difference between neck collar and neck harness must have another reason. Neck collars do cause much more and more severe wounds on the neck of the sows than neck harnesses. Any traction on these wounds will cause pain. To avoid traction the sow will change her way of getting up. This in its turn causes a higher number of lesions on the udder. Here the environment causes a change in health state (wounds in the neck); this changes the behaviour, and results in an increase in the level of lesions of the integument.

Another way the environment is influencing the lesion pattern is by inducing changes in behaviour via a psychical influence.

160

3. Health

A third major factor influencing the state of the integument is the health of the individual. An external insult will cause lesions of the integument. The nature of the lesions will depend upon the nature of the insult but also upon the resistance of the integument. This resistance is directly influences by the state of health of the individual. General disease weakens the resistance of the integument, and the accompanying drop in condition will lead to a smaller layer of tissue between protruding bones and the skin. These factors have a negative influence on the occurrence of lesions of the integument.

Deviations in health also have an influence on the psyche, and result in changed behaviour. This again will result in changes in the lesion pattern of the integument. Could the difference between neck collar and neck harness in figure 3 already be explained by health differences, in figure 4 another example of the influence of health on the pattern of the lesions of the integument is given.

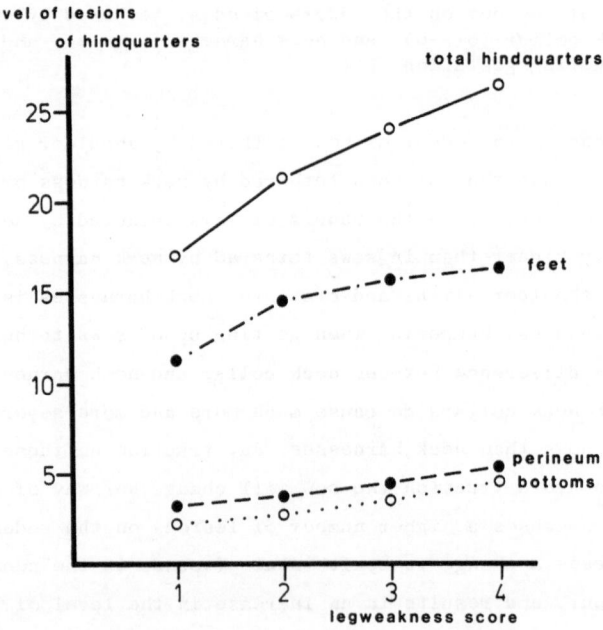

Fig. 4 The level of lesions of bottoms, perineum area, feet and total hindquarters of sows with different leg weakness scores (1 = good ....
4 = poor leg soundness).

161

Bottoms, perineum area, as well as hindfeet show an increase of lesions, when the leg soundness becomes worse. The more leg weakness is pronounced the more it causes deviations in behaviour. Presumably also the resistance of the integument is negatively affected when leg weakness becomes worse. Both factors will be responsible for the increase of the level of lesions of the hindquarters.

DISCUSSION AND CONCLUSIONS

As the literature shows, a systematic inspection of the integument is not commonly used in animal welfare research. When inspection is carried out, lesions of the integument are considered to be pathological deviations from the optimal health state, and therefore indications of reduced well-being (Algers, 1980; Bäckström, 1973; Lindquist, 1974; Simonsen, 1980, Troxler, 1981). Part of the lesions are attributed to behaviour, but generally no further conclusions are drawn from that assumption. On two occasions the presumption is put forward, that, except for the direct pathological influence a higher level of lesions might coincide with a decreased well-being (van Veen, 1981; Vellenga, 1980).

In this paper it is shown that behaviour, health as well as the environment affect the lesion pattern of the integument. The influence can be direct or indirect through a combination of these factors. In figure 5 environment, health and behaviour are put into a diagram, in which the influences on each other and on the state of the integument are represented by arrows.

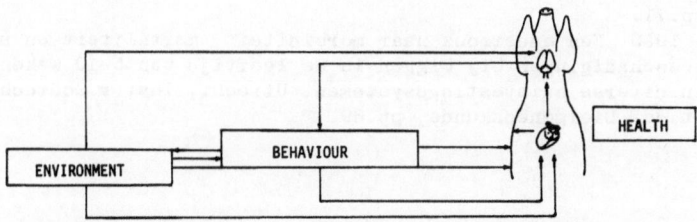

Fig. 5 Diagram of influences of environment, behaviour and health on each other, and on the state of the integument, represented by arrows.

This total complex of influences on the integument includes a behavioural part and a health part. Behaviour and health are both prime indicators of well-being and do reflect in the integument. Therefore the state of the

integument might very well be a good indicator for well-being of the individual animal. Major deviations in behaviour and health will show up as
changes in the lesion pattern. Minor changes both in health and in behaviour will not always change the state of the integument. This has consequences for any method which determines well-being on the bases of lesions of the integument alone, and therefore also for the Ekesbo Method.

In conclusion it can be said that the state of the integument is a
reflection of the well-being of the individual. It is likely that a method
based on lesions of the integument can distinguish between two environments
when differences are big enough, but minor differences might be overlooked.

REFERENCES

Algers, B., 1980. Utredning av burhållning ao avvanda smågzisar i åldern
    3-8 weckor uo djurkålso-och djurskyddsynpunkt. Sveriges Lantbrukuni-
    versitet, Skara.
Bäckström, L., 1973. Environment and health in piglet production; a field
    study of incidences and correlations. Diss., Stockholm, Acta Vet.
    Scand. suppl. 41, pp.240.
Lindquist, J.O., 1974. Animal health and environment in the production of
    fattening pigs. Stockholm. Acta Vet.Scand. suppl. 51, pp.78.
Lorz, A., 1973. Tierschutzgesetz, Kommentar. München, Verl. C.H. Beck,
    pp.70.
Troxler, J., 1981. Beurteilung zweier Haltungssysteme für Absetzferkel.
    In: Aktuelle Arbeiten zur Artgemässen Tierhaltung 1980. KTBL.Schrift
    264. Darmstadt-Hiltrup, 1981.
Simonsen, H.B., Vestergaard, K. and Willenberg, P., 1980. Effect of floor
    type and density on the integument of egg-layers. Poultry Sci., 59,
    2202-2206.
Veen, H.M. van, 1981. Verslag van een onderzoek naar morbiditeit, mortali-
    teit en uitwendige beschadigingen bij biggen in twee typen kraam-
    opfokhokken. Utrecht. Inst.v.Zoötechniek, Faculteit der Diergenees-
    kunde, pp.71.
Vellenga, L., 1980. Een onderzoek naar morbiditeit, mortaliteit en uit-
    wendige beschadigingen bij biggen in de leeftijd van 5-10 weken, ge-
    houden in diverse huisvestingssystemen. Utrecht, Inst.v.Zoötechniek,
    Faculteit der Diergeneeskunde, pp.69.

DISCUSSION

Chairman: R. Moss (U.K.)
Many questions were directed to all the speakers with interest being shown in the studies comparing calves and 2 year old fattening animals kept on slats with those kept on straw.

The extensive and in depth studies of pig production systems undertaken by J.P. Tillon/France were welcomed by a number of speakers but it was noted that despite the problems encountered, producers very rarely took the opportunity to avail themselves of advice from specialists.

The paper on abomasal ulceration raised a number of questions. K. Dämmrich/FRG revealed that no clinical symptoms whatsoever were seen in calves with the lesions.

The chairman suggested that the papers presented in the session indicated that animal welfare could be vastly improved if the amount of present knowledge were applied without recourse to consideration of so-called behavioural needs. He reminded the seminar of the important concept of zoo-technical design of equipment and housing being pursued at the Scottish Farm Building Research Institute. Finally, the chairman wondered if perhaps we may sometime in the future consider accepting some degree of traumatic injury if, by doing so, livestock were thus provided with other more essential behavioural needs. To accept such injuries, however, these needs must be acceptably defined. An example of such a situation could be the need for freedom of movement and social contact which, on the other hand, could lead to increased aggression and injury. Good health is a prerequisite of animal welfare and the veterinary profession has an essential role to play in achieving this goal.

SESSION II

SIGNIFICANCE OF INDICATORS RELEVANT TO ANIMAL WELFARE
Production Performance

Chairman: J. Langholz

SIGNIFICANCE OF PRODUCTION PERFORMANCE TRAITS
AS INDICATORS OF ANIMAL WELFARE

J.M. Bienfait, B. Nicks, C. Van Eenaeme
Faculté de Médecine Vétérinaire de
l'Université de Liège
Rue des Vétérinaires, 45
1070 Bruxelles - Belgium

ABSTRACT

The first aim of domestication of farm animals is to control and increase the animal performance traits in order to satisfy economically the
requirements of man for food with great biological value. The profit of
animal production depends on economical factors and performance traits :
the former are not related to animal welfare but the latter are influenced
by all the constraints imposed to the animals. The literature as the clinical observations show that animals are very sensitive to modifications
in the conditions of management, micro-climate and indoor equipment of the
buildings. The purpose of improvement of farm conditions is to tend to the
best economical arrangement between animal welfare and stockman welfare.

INTRODUCTION

The first aim of farm animals productions is to provide mankind with
food of high biological value, featuring high contents in energy, protein,
amino acids, vitamins, minerals and trace elements.

The second aim is to satisfy the demand of man for animal food in preference to vegetable food, generally in excess to his specific biological and
physiological needs : a phenomenon in German language called "Luxus Konsum"
and which is responsible for physiological disturbances such as obesity,
digestive pathology, cardiopathy, etc.

In order to achieve these goals, man has domesticated farm animals by
controlling *their genetics* by selection and may be to morrow by genetic
engineering, *their reproduction* by manipulation of the sexual cycle, artificial insemination and to morrow by embryon transplantation, *their nutri-
tive requirements* by adequate feeding, *their housing* by assembling animals
in specialised production plants.

All these interventions are only possible as a result of the highly developed adaptive power of farm animals to constraints to which they are exposed, especially with regard to their sexual, feeding, locomotive and social
behaviour.

The pathology of the wild animal has been replaced by the pathology of
domestication obliging veterinarians to a constant re-evaluation of etiolo-

gy, symptomatology, prophylaxis and to preventive rather than curative
treatments.

In nature, animals dispose only of bedding sites, they are submitted to
the struggle for life and to the severity of the climatic conditions; they
must cope with seasonal variations of feed avaibility; their reproduction
is essentially governed by instinct and by limitations of local populations and consequently their productivity is low. Domestic animals have
attained a remarkable productivity, put in cóncrete form by the rentability of their productions.

Finally the rentability of animal productions is the reason of pursuit of
the age-long experiment of domestication; it is indeed clear that man has
become a stockman in order to make some benefits; otherwise he would have
remained a hunter.

THE PRODUCTION PERFORMANCE TRAITS

The economy of a production is influenced by two types of factors:the
"economical factors" and the "zootechnical determinants". The economical
factors vary in function of market prices of animals, feedstuffs, manpower
and money rates. These factors are in no way related to animal welfare.
The zootechnical determinants of production costs measure animal performances in function of species, breed, feeding and farming conditions.
These factors are biological parameters which are easily quantified by
geneticians, nutritionists and hygienists when interpreting results of
their experiments or of their observations in animal farms. Among these
zootechnical determinants some can be defined as *primary determinants* :
they are the result of the direct recording of animal productions in absolute units; some could be labeled *relative* e.g. the productions per time
unit or per kg live weight. The latter result from combining two or more
primary determinants. Other express the performances by incorporating factors of survival or of quality of productions, e.g. liters milk sold per
cow, amount of eggs sold per hen initially present, number of weaned piglets per sow-place in the farm : they could be called *real determinants*.
Each type of determinant is used in function of the aim of the study :
with regard to animal welfare the real determinants have the greatest discrimination power because they include all risks and hazards of stock farming comprising diseases and death in real world farming conditions and
not in strictly controlled experimental station conditions.

For instance, the following formula explains a method of calculation of

the total net benefit (BNT) obtainable by beef cattle during the growth
fattening period (Bienfait et al.,1978) :

$$BNT = GT.V - (Qb.B + Qc.C) - T.D + Pi (V - A)$$

In the above cited formula the economical factors are : V and A, respectively the sales and purchase prices per kg live weight, B, C and D, the
unit prices of basic feed, supplement feed (concentrates) and of daily
exploitation costs per animal, e.g. manpower, capital investment, interests and writing downs, insurance policies, veterinarian fees and general exploitation costs.

The primary zootechnical determinants are : the total weight gain GT obtainable from an initial weight Pi, during a time T, with a feed intake
of basic feed Qb and of supplement feed Qc.

Per kg weight gain the total net benefit is :

$$BNT/kg \text{ weight gain} = V - (Ib.B + Ic.C) - \frac{1}{GQM}.D + \frac{Pi}{GT} (V - A)$$

This formula introduces the *relative zootechnical determinants* such as
feed conversion rates of basic and supplement feed (Ib = kg basic feed/kg
weight gain and Ic = kg supplement feed/kg weight gain), the reciprocal
daily weight gain 1/GQM which indicates that the contribution of the daily
non feeding costfactors is inversely related to the growth rate, and finally, the determinant Pi/GT which measures the difference between sales
and purchase prices per kg weight gain.

In fact, the real net benefit per animal should be formulated as :

$$BNR = S.GT.V - (Qb.B + Qc.C) - T.D + Pi (V - A)$$

where S stands for survival rate of the animals on this type of fattening
treatment.

It can be equal to 1 if all animals reach their slaughter weight, and it
would be 0.98 if for instance 2% of the animals die before the end of the
fattening period. Consequently this term measures the hazards burdening
the real benefit of this production.

These formulae indicate that for an identical feeding ration and a given
type of animals and management, the zootechnical determinants are cons-
tants, fixing the benefits as a function of the economical factors.

The improvements in nutrition, genetics or management of the animals can
have an "in se" effect on the benefits if these improvements result in a
better quality of the final product in such a way that the unitary sales
price is affected. If the variations of the market prices with equal pro-

duct quality conditions have nothing to do with animal welfare, each intervention affecting the animal and consequently the zootechnical determinants
is not indifferent to animal welfare.

From a *genetics point of view*, a cow needs only a production of 600 to 700
liters milk per lactation period in order to secure her reproduction and
to suckle her calf. It is difficult to estimate if cow welfare has been
improved when selection has rised her genetic production potential to
7,000 or more liters milk. The same conclusion applies to hens with egg
production rates of 280 to 300 eggs per year.

From a *reproduction point of view*, the speeding up of the pregnancy-lacta-
tion cycle of the sow is certainly not indifferent neither to sow welfare
nor to piglet welfare to the point to which exaggeration in this field
results in fertility problems for the sow and in behaviour troubles for
the piglets which could worsen their ulterior performances.

When the *nutritionist* has to elaborate a ration for a cow with a milk production of 7,000 l, he undoubtedly makes a contribution to cow welfare if
the resulting ration entirely meets the cow's requirements in relation to
her genetical potential for milk production and reproduction, thereby pre-
venting the occurence of deficiency diseases which not only decrease her
performances but could even put her life in danger.

The main targets of "animal welfarists" are however the *manager and
hygienist* in charge of production techniques and animal housing.

The evolution of the economy of animal productions during the last decades,
imposed to farmers scale expansion,increased mechanization and specialisa-
tion in order to survive. Factors in favor of scale expansion of production farms are : the low benefit per animal and the decrease of building
and equipment costs per animal.

In order to lower manpower expenses, the rudest duties such as feed distribution, production harvesting and cleaning of the installations have
been mechanized.In a few years, considerable improvements have been made
in order to alleviate and reward the work of manpower : a non productive
work load has been replaced by directly productive activities : hygienic
control and administration activities of the production.

In bovine productions the most striking fact in the evolution of housing
conditions during the last years is the development of the loose housing
system. Frison (1977-1980) mentions that in Holland, Scotland and Ireland,
95% of new constructions for cows are of the cubicle type while this
amounts to 40% in England and 30% in France. In Belgium the proportion of

cubicle loose housing in new constructions reaches 24% (Martens et al.,
1980).

The success of cubicle loose housing in opposition to traditional cowsheds
is mainly due to the lowering of manpower requirements. Grillot cited by
Carrotte (1981) reports minimum labour requirements of 11.2 working minutes per cow and per day for a cowhouse of 50 cows with the cowshed system
and 6.9 minutes with cubicle loose housing.

In pig production, the most striking facts in the evolution of housing are :
the individual confinement of sows and the utilisation of slatted floors.
In Belgium, 68% of new constructions (Stat. 1977-1978) for empty and pregnant sows have individual boxes and 30% have a tether device. In 90% of
them partial slatted floor is provided on the posterior side (Ministry
of Agriculture, 1980). In Great Britain in large pig production units 75%
of the pregnant sows are kept individually and 1 out of 4 is tethered
(Baxter, 1981). The suckling sows are equally more and more tethered. In
Belgium, 1 out of 2 in new constructions and 75% of the stalls have a partial slatted floor. The latter allows to decrease the cleaning time per
day and per litter from 5.5 minutes (concrete floor) to 0.92 minutes
(Mc Fate et al., cited by J. Le Dividich, 1970). In fattening swine plants,
only 1% of the new constructions in Belgium are planned with a concrete
floor (Ministry of Agriculture, 1980).

In the poultry sector, the battery cage system for laying hens is widely
used in the E.E.C. so that about 80% of the 280 millions laying hens of the
Community are reared in cage (E.E.C. Document, 1981).

These changes in housing systems of the domesticated animals were accompanied by a search for better surrounding conditions in order to allow a
better expression of the animal's production potential. In some cases, the
farming technique was changed simultaneously. The "all in-all out" technique was developped in different areas of animal production. Its application depends on a better setting up of the housing. The criteria which
were taken into consideration in the search for better management of animals
and of better surrounding conditions and equipments were production criteria. Animal performances were evaluated against the costs of these performances. The modifications yielding the best results were retained.

In view of the recent evolution of stock breeding the question rises if,
in a certain sense, animal welfare has not been overlooked. This rises the
question about the meaning of the relationships between productivity and
welfare.

MANAGEMENT, PERFORMANCE TRAITS AND WELFARE

The study of performance traits for different types of housings has
shown the advantages of some management schemes as the all in-all out
techniques and the production of specific-pathogen-free (SPF) animals. Table 1 shows the results obtained in farms by Debruyckere and Martens
(1970) when changing from a conventional sow farrowing unit to a unit sub-
divised in smaller compartments. A simple analysis of piglet mortality
rate reveals a net improvement of the health status of the herd and consequently its welfare, health being one of the essential conditions for welfare. The joint examination of these performance traits between scientist
and farmer represents a concrete and efficient way to illustrate the usefulness of new management techniques improving simultaneously productivity
and welfare.

The all in-all out technique with smaller units has also proven its usefulness in pig fattening (table 2). For reasons of improvement of performance traits, Daelemans et al. (1980) recommend compartimentation for new
buildings in spite of significantly higher costs.

Similarly the study of the production parameters gives the best illustration of the advantages of the SPF method in pig fattening. Table 3 compares growth rate and feed conversion of fattening pigs before and after the
establishment of a SPF scheme and also between herds with different health
status. These improvements resulting from a simple change in the management are an indication of an increase of health status of the herd and
consequently of its welfare.

INDOOR ENVIRONMENT, PERFORMANCE TRAITS AND WELFARE

The total environment of an animal is highly complex and it is probably impossible to factorize it completely in any useful way. Generally,
the environmental factors are divided into two groups : those which are
primarily thermal in their impact on the organism and those which are not.
The thermal factors include air and mean radiant temperatures, air speed,
humidity and the physical nature of the floor. The non thermal factors include light, sound, presence or absence of other animals, etc.

The perception of a thermal comfort in a given environment is certainly
one of the components of welfare. It is also a component of animal productivity. Attemps have been made to set relationships between thermal comfort and performance traits. The poorer the capacity of adaptation of the
animal to climatic stress e.g. the new-born animals, the closer are these

relationships.

From a theoretical point of view, homeotherms reach their maximal thermal comfort within the zone of thermoneutrality. It is also in this range of temperatures that feed conversion defined as metabolisable energy intake per unit of energy gain is maximal. A maximal productivity is associated with an optimal thermal comfort.

From a practical point of view, relationships between climatic parameters and performance traits have been proposed. For dairy cows in a cold environment, a simple model based on field results in Canada, has indicated that milk production of cows receiving adequate diets declines at a rate of 0.25 kg/cow for each 10°C decrease in average daily temperature below 5°C (Christison, 1978). Figure 1 shows response functions developed for lactating dairy cows in hot weather conditions. This figure represents the expected milk production, conception rate, rectal temperature and hay intake for an "average" cow as a function of a Temperature-Humidity Index (THI).

In the fattening of pigs, Verstegen et al. (1978) have analyzed the results of growing an finishing trials from literature to set relationships between several performance characteristics (rate of gain, feed conversion, feed intake and slaughter quality) and temperature. Figure 2 shows fattening and carcass traits of pigs housed at various ambient temperatures expressed as a percentage of the 15°C values. The rate of change of feed intake, rate of gain and feed conversion were respectively : -19.5 g/day, 8.1 g/day and -0.033 kg/kg per °C increase in temperature between 5 and 20°C. At computed similar intakes of feed, rate of gain decreased with 15 g/°C below 20°C. Backfat thickness was as a mean decreased in the cold as was lean meat to fat ratio. From these data, the optimal temperature for animal feed conversion is about 20°C. De la Farge (1976) proposed a bioclimatic index (BI) to give an assessment of the thermal comfort feeling of fattening pigs :

$$BI = 0.89\ t_{gt} + 0.05\ RH - 1.81\ s + 0.02\ w - 21.15$$

where t_{gt} = globe temperature (°C), RH = relative humidity, s = air speed (m/s) and w = liveweight of pigs (kg).

Animals will be comfortable for a value of BI near 0. The results of Texier et all. (1979) showed that a BI-value in the range of 0 and +4 is recommended to obtain maximal performances. The accuracy of the predictions of such an index are still limited in practice on a widespread basis but improvements are possible.

Response functions have also been developed for other species and other types of productions. For instance, Sainsbury (1980) reported a decrease in the production of laying hens of one egg per hen per year per °C below 21°C.

The influences of non-thermal factors of the environment on welfare and productivity are generally not so well defined as the effects of thermal comfort. For instance less experimental information is available on the effect of manure gases, odor, aerial dust and microbial air content on animal production and welfare, but clinical observations in farms indicate that these factors can affect substantially animal health status.

INDOOR EQUIPMENTS, PERFORMANCE TRAITS AND WELFARE

It is well established that animal performances and welfare are strongly influenced by the components of the building. The characteristics of the floor (thermal insulation, softness, area per animal) affect both performance traits and comfort.

Considering the thermal properties of different types of floor, Bruce (1979) developed a relationship between feed intake, air temperature and floor type. Using this model, it has been shown that for pregnant sows kept in a building with a temperature of 13°C, the maintenance ration required is 2.5 kg of meal per day with a concrete floor and 2.0 kg with straw.

Most of the space requirements (m^2/animal) usually proposed nowadays are based on studies with performance traits. For the different types of floors there is a minimal area per animal below which productions decrease. This area depends on the number of animals in the group. Without going as far as pretending that the minimal area allowing full expression of the animal's productivity potential, corresponds to a minimal area for welfare, this minimum indicates nevertheless a breakdown point with regard to animal welfare. It is noteworthy that an increase of comfort is not necessarily correlated to an increase in area. For instance with laying hens agressivity increased with stocking rate between 400 to 800 cm^2 per hen in cage (E.E.C. Report, 1981). Similar observations are made with alternative sow housing systems (English et al., 1982). Moreover, an increase in the volume of the buildings makes climatization more difficult and expensive.

In the choice of floor type tendency has been to recommend the easiest and the cheapest solution. At the present time however more and more attention is paid to animal requirements as an answer to their decrease in productivity due to housing conditions. For instance, in pig production, fattening performances (growth rate, food conversion) on total slatted floors are worse than those on partial slatted floors (Maton et al., 1978). On this base, Daelemans et al. (1980) recommend buildings with partial slatted floors in spite of a higher cost ($\pm$ 8.5% in a piggery with 500 places). If animal welfare is better on partial slatted floors than on total slatted ones, it can be said that taking into account the performance traits has improved animal welfare. Finally, for obtaining the best performances the use of straw is still better than the partial slatted floor but this solution is not retained because the use of straw is very expensive and it represents an important increase of a laborious piece of work. Undoubtfully there will be improvements of floor quality in the near future. Research of new floor materials for better performance traits will improve animal welfare.

The equipment of farm buildings was also planned in order to enhance animal productivity, at the same time alleviating the farmer 's work. These considerations explain the success of the battery cage system for hens. However does this general and total improvement of productivity also result in an improvement in welfare? With our present-day knowledge it is difficult to give a clear-cut and definitive answer to this question, for, if productivity is easily quantified, animal welfare is not. However, zootechnical performances indicate that certain components of animal welfare such as health status, thermal comfort and quality of production were improved in a general way.

The question can be rised if sometimes the improvement of productivity and general welfare of the collectivity is not obtained to the detriment of some individuals. In pig production, the caging or tethering of sows reduced piglet losses (table 4). On one hand the freedom of the sows has been limited, which is a negative aspect of welfare, and on the other hand, the piglets were better protected against crushing, which is a positive effect of welfare.

When analyzing animal performances in apparently similar conditions, i.e. same types of animals, feeding and housing systems, substantial differences are occasionally observed between farms. Experts impute these discrepancies to differences between farmers in professional skill, sense of observation, technical knowledge and ability to correctly apply this knowledge. It is highly probable that in similar exploitation conditions a better stockmanship results in improvements of both productivity and animal welfare.

CONCLUSIONS

Adequate control of the biology of domesticated animals results in constant improvements of management methods in specialized buildings.Domesticated animals live 24 hours a day, during several months and even their entire life-time in the same place. Consequently they are "biological reagents" of very high sensitivity, which will manifest very rapidly their uncomfort with regard to their behaviour, productivity and health. One could easily agree with Curtis (1972) as he states "uncomfortable animals usually perform suboptimally".

The objectives of housing and equipment are :

For the animals

1. To establish a microclimate in favour of their health and productivity;
2. to promote a housing microflora with an increased resistance to pathogenic microorganisms;
3. to protect them against predators;
4. to provide them with a feeding equipment appropriate to their number and feeding behaviour;
5. to furnish adequate facilities protecting them from injuries and abnormal behaviour;
6. to allow them to achieve more plentiful, healthier and higher quality productions.

For man

1. To provide staff in charge of maintenance of the animals with an efficient tool for feed distribution, harvesting of productions, manure disposal and animal's handling;
2. to create climatic and hygienic working conditions preserving his health, alleviating his work and contributing to his social promotion;
3. to provide him with a production unit suited to his skill and his physical and moral abilities for the management and direction of the herd.

For the environment

To secure an adequate protection against animal pollution which rises in an alarming way with increasing animal concentrations.

The best farms are those which assume the best compromise of all these constraints, in order to minimize production costs and to optimize the profit of the production. Thereby the product sales price should be compatible with an adequate availability of animal food for the less favoured classes of human population. "Animal welfare" must not come into competition with "human welfare".

REFERENCES

Aumaître, A. 1977. Des maternités bien aménagées réduisent la mortalité des porcelets. L'Elevage porçin, n° 67, 39-43.

Baxter, S.H. 1981. Welfare and the housing of the sow and suckling pigs. In "The welfare of pigs" (Ed. W. Sybesma). (Martinus Nijhoff, The Hague) pp. 276-311.

Bakx, J. e.a. 1979. Onderzoek op varkensmestbedrijven van integratiegroepen. Bedrijfsontwikkeling, 10, 375-388.

Bienfait, J.M., Van Eenaeme, C., Lambot, O., Pondant, A. and Denayer, J. 1976. Une nouvelle méthode d'appréciation zootechnique et économique des régimes distribués aux animaux destinés à la production de viande. Revue de l'Agriculture, 29, 289-309.

Bruce, J.M. 1979. Heat loss from animals to floor. Farm Building Progress, 55, 1-4.

Carrotte, G. 1981. In : Logement des vaches laitières. La France Agricole, 4 dec.

Christison, G.I. 1978. Weather and dairy cows. Confinement, 3, 6.

Commission des Communautés Européennes, 1981. L'élevage des poules pondeuses en cages : aspects économiques. Doc. Sec.(81) 1283.

Curtis, S.E. 1972. Air environment and animal performance. J. Anim. Sci., 35, 628-634.

Daelemans, J., Martens, L. and Maton, A. 1980. Les investissements dans les porcheries d'engraissement et leur équipement. Revue de l'Agriculture, 33, 483-513.

Debruyckere, M. and Martens, L. 1980. La rentabilité de la mécanisation accrue dans l'élevage des truies. Revue de l'Agriculture, 33, 463-481.

de La Farge, B. 1976. Le porc à l'engrais et son habitat. Bulletin de l'Institut Technique du Porc, 4, 59-71.

English, P.R., Baxter, S.H. and Smith, W.J. 1982. Accomodating in the welfare dimension in future systems. British Society of Animal Production, winter meeting 1982, paper 26.

Frison, M. 1979. Bâtiments d'élevage bovin : tendances en France et à l'étranger. In "Bâtiments d'élevage" (Ed. Association Française du Génie Rural, 30 rue Las Cases, 75340 Paris Cedex 07)

Frison, M. 1980. Des bâtiments laitiers mieux adaptés aux progrès. L'Elevage bovin, n° 104, 75-79.

Hahn, G.L. 1981. Housing and management to reduce climatic impact on livestock. J. Anim. Sci., 52, 175-186.

Le Dividich, J. 1976. Recherche des conditions optimums de milieu pour l'élevage du porcelet avant sevrage. In "Bulletin de l'Institut Technique du Porc" pp. 31-43.

Martens, L., Daelemans, J., Maton, A. 1980. Les investissements dans les étables pour vaches laitières et leur équipement. Revue de l'Agriculture, 33, 433-450.

Martineau, G.P., Broes, A and Martineau-Doize, B. 1982. Intérêts et limites de l'assainissement des élevages dans l'économie de la production porcine. Ann. Méd. Vét., 126, 279-313.

178

Maton, A., Daelemans, J. and Lambrecht, J. 1978. Le revêtement du sol des loges pour porcs de boucherie : ses propriétés techniques et ses effets zootechniques. Revue de l'Agriculture, 31, 308-317.

Mc Fate, K.L., Veum, T., Sprouse, W. and Sandidge, R. 1974. Proceed International Livestock Environment Symposium. pp. 331-337.

Ministère de l'Agriculture, Direction du Génie Rural, Bruxelles, 1980. Les tendances en matière de constructions des bâtiments d'élevage bovin et porcin. Revue de l'Agriculture, 33, 401-432.

Sainsbury, D. 1980. Poultry health and management. (Ed. Granada, London).

Texier, C., de La Farge, B. and Granier, R. 1979. Influence des variations des principaux facteurs de l'ambiance en porcherie d'engraissement. Journées Rech. Porcine en France, 153-164.

Verstegen, M.W.A., Brascamp, E.W. and Van Der Hel, W. 1978. Growing and fattening of pigs in relation to temperature of housing and feeding level. Can. J. Anim. Sci., 58, 1-13.

TABLE 1   Piglet mortality before and after compartimentation of the farrowing unit.

	Number of sows in the farm		Mortality from birth to weaning (%)	
	Before compartimentation	After compartimentation	Before compartimentation	After compartimentation
Farm 1	200	200	12.3	8.3
Farm 2	50	70	30.6 and 29.8(1)	18.5 and 14.1(1)
Farm 3	-	-	28.6 and 20.6(1)	15.0 and 18.5(1)

(1) : Two successive years.

(Debruyckere and Martens, 1980)

TABLE 2   Relationship between size of units and the growth rate of fattening pigs.

Number of places per unit	Growth rate g/day
<100	680
100 - 150	668
>150	635

(Bakx, 1979)

TABLE 3   Performance traits of "conventional" and "SPF" fattening pigs.

| | | | Pigs | |
			Conventional	"SPF"
Growth	Betts	1955	417 to 510	540 to 590
rate	Shuman	1956	653	689
(g/day)	Young	1959	644 to 712	717 to 862
	Peo	1965	650 to 765	715 to 806
	Sharman	1971	510	582
	Sickel	1971	591	635
	Forest	1978	580	640
Feed	Peo	1965	3.29 to 3.44	2.88 to 3.09
conversion	Heard	1967	3.95	2.80
(kg/kg)	Keller	1971	3.80 to 4.00	3.00 to 3.10
	Sharman	1971	3.73	3.19
	Sickel	1971	3.65	3.25
	Forest	1978	3.45	3.10

(Martineau et al., 1982)

TABLE 4   Piglet losses with different types of equipment in the farrowing pen.

| | Equipment of the farrowing pen | | | |
Mean number of piglets lost (per litter)	Without piglet protection	Rails of protection	With crate	Tethered sows
Birth - 48 h	1.40	1.37	1.25	1.24
Birth - weaning	2.28	2.30	2.17	2.03

Source : Aumaître and Dagorn (1975)
n = 230000 litters
(Aumaître, 1977).

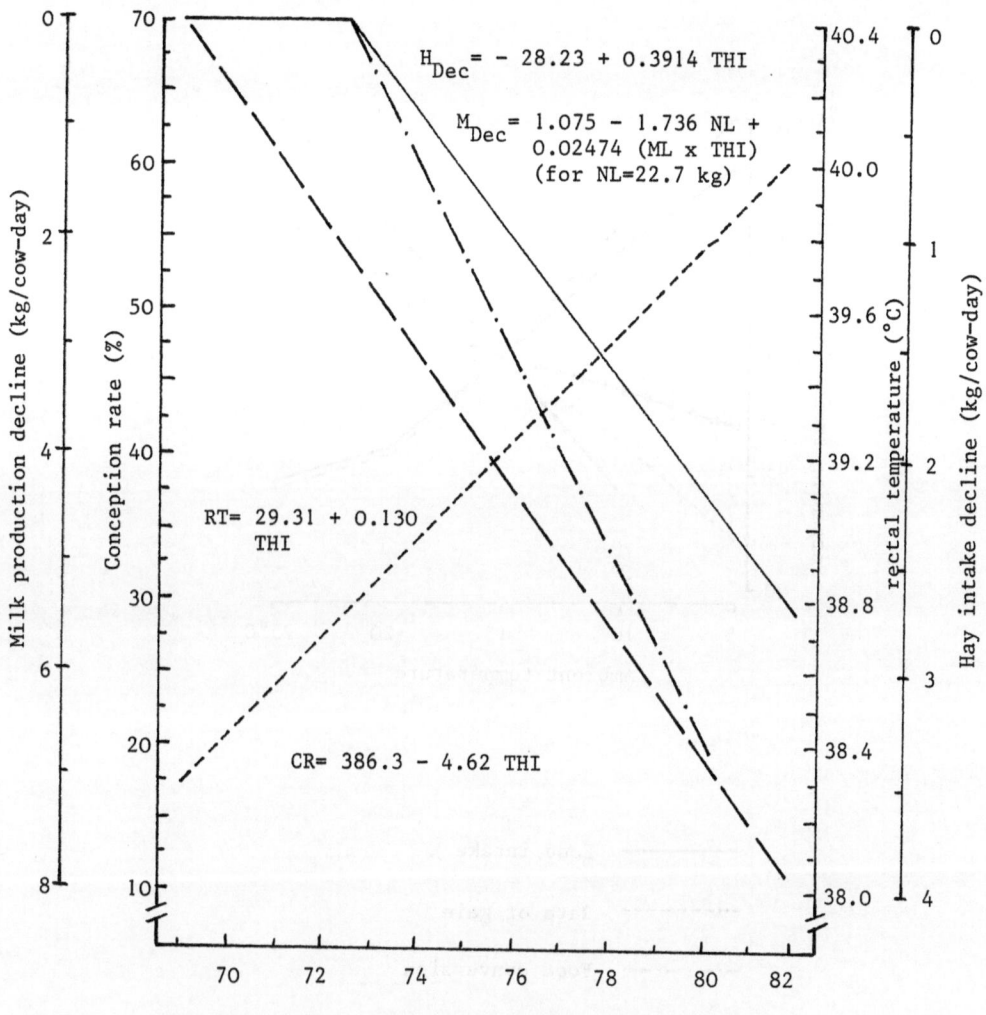

FIGURE 1
Milk production decline (MDec), hay intake decline (HDec), rectal temperature
(RT) and conception rate (CR) responses of lactating dairy cows shaded but
exposed to hot weather, to a temperature-humidity index.

THI = t_{db} + .36 t_{dp} + 41.2 where t_{db} = dry-bulb temperature (°C) and t_{dp} =
dew-point temperature (°C).

NL refers to the normal level production of the cows at THI   70 the response
functions MDec (Berry et al., 1964) and HDec (Osburn and Hahn, 1968) are based
on laboratory data that were field valided, whereas those for RT and CR were
derived from field data (Ingraham, 1974).

(Adapted from Hahn, 1981).

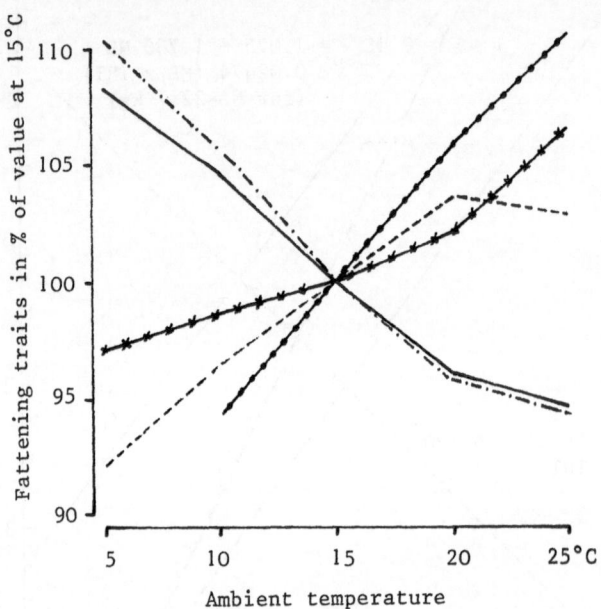

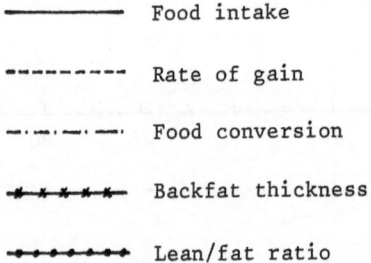

——————— Food intake

— — — — — — Rate of gain

—··—··—··— Food conversion

—✶—✶—✶— Backfat thickness

—•—•—•—•— Lean/fat ratio

FIGURE 2   Fattening and carcass traits of pigs housed at various ambient
temperatures and expressed as a percentage of that trait at 15°C.

(Verstegen et al., 1978).

FARROWING ACCOMODATION AND LOSSES, VITALITY AND DAILY GAIN OF PIGLETS AS INDICATORS OF WELL BEING

H. BEKAERT

National Institute of Animal Nutrition
B-9231   Melle, Belgium

ABSTRACT

In a study piglet losses and vitality of piglets, as measured by daily gain, as indicators of well being, were investigated, using 2 farrowing facilities, namely standard farrowing pens and farrowing pens equiped with farrowing boxes. A total of 90 sows and 782 piglets were involved. The experimental results do not reveal any improvement in well being of the piglets as measured by mortality rate and daily gain due to a modification of the farrowing pen.

INTRODUCTION

One of the most critical periods in the piglets life is undoubtedly the period of birth. Indeed, in this period the piglet is faced to different environmental changes. It is generally known that the highest mortality rate is noted immediately after birth (Braude, 1972 ; English et al., 1977 ; Dreihsig, 1978 ; Hughes and Varley, 1980). This means that a maximum of precautions are necessary to prevent this high mortality rate, avoiding by these important financial losses and improving the well being of the piglet. A reduction of mortality can be realized by creating an optimum microclimate for the newborn piglets and by using farrowing crates (Aumaître, 1977). Indeed one of the principal causes of death is overlying (English et al., 1977 ; Dreihsig, 1978 ; Hughes and Varley, 1980). Some years ago a reduction in piglet losses of 7.3 % was obtained as a result of the introduction of farrowing crates. The question arises if other precautions can be taken to reduce further mortality and if it is possible to improve as such the welfare of the piglet by equiping the farrowing pens with farrowing boxes and if this equipment has an influence on the vitality and daily gain from birth to weaning age.

Communication R.V.V. n° 514.

MATERIALS   AND   METHOD

To study the effect of farrowing accomodation on the well
being of the baby pigs, mortality, vitality and daily gain were
compared of the piglets born in farrowing conditions with farrowing boxes or in farrowing conditions without these boxes.
By installing the boxes at the expected farrowing day, the pens
were adapted as follows : 3 metal slats behind the sow were inclined so that they formed a chute, ending in the farrowing box
bedded with straw and equiped with an electric heater. Heat was
delivered by 2 lamps of 250 watt, one in the creep area and
another one above the box or behind the sow by lack of farrowing boxes.

In this trial 41 sows of the Belgian Landrace farrowed in
pens with diagonally placed crates and 49 sows farrowed in
identical pens provided with farrowing boxes. The number of
piglets concerned was respectively 367 and 415.

RESULTS
1. Post-natal mortality

From birth to weaning at $\pm$ 28 days the post-natal mortality decreased from 18.3 % to 15.9 % by using a farrowing
box. It is generally accepted that most of the casualties
occur during the first week. By then, the effect of the farrowing accomodation should be shown by the mortality rates
during the first days. An analysis of the mortality rate in
relation to farrowing accomodation and age is mentionned in
table 1.

Considering the first two days, one could be tempted to
derive that the use of a farrowing box decreases the mortality rate, but from the 3rd day on, this mortality rate be-
cames higher. Moreover, it is worth mentioning that the mortality in the first 7 days amounted to 72.8 % of total
deaths with the farrowing box and 64.2 % without that box.

It is also interesting to know the principal causes of
death. Contrary to the expectations 55.6 % of the died piglets were crushed on birthday by using the farrowing box
against 52.9 % without the box. In general, during the total

TABLE 1   Piglet mortality, related to farrowing accomodation and age

Age (days)	1	2	3	4	5	6	7	First week
With box	37.5	12.5	27.1	10.4	4.2	2.1	6.2	72.8
Average birthweight of the died piglets (kg)	1.05	0.78	1.05	1.00	1.15	1.30	0.90	
Without box	39.5	18.6	16.3	9.3	2.3	2.3	11.6	64.2
Average birthweight of the died piglets (kg)	1.05	1.11	0.94	0.94	0.90	1.70	1.24	

pre-weaning period, death was caused in 15.2 % by crushing when using the farrowing box, against 13.4 % without this box. This leads to the conclusion that the farrowing box does not prevent crushing. This results are conflicting with those of Robertson and McCartney (1980) who obtained a reduction in mortality due to crushing from 64% to 27 % by using farrowing boxes.

Further on death was for 66.7 % caused by starvation or illness when using the farrowing box against 62.7 % without the box. This may indicate that the slope that the piglets have to overcome in order to reach the udder is substantial, leading to a to late and a to low intake of colostrum. From this point of view the use of a farrowing box cannot be recommended.

A better micro-climate and by then the farrowing box would protect perhaps better the lightest piglets. In order to study this, mortality as related to birthweight is given in table 2.

Out of table 2 one may conclude that the farrowing box does not protect better the lightest piglets ; for survival chances a higher birthweight is of more importance than the use of the farrowing box.

TABLE 2   Mortality rate as related to birthweight
            and farrowing accomodation

Birth-weight (kg)	Mortality (%)	
	With farrowing box	Without farrowing box
0.8	68.4	80.0
0.8 - 0.9	41.7	30.4
1.0 - 1.1	20.8	27.4
1.2 - 1.3	15.2	14.3
1.4 - 1.5	11.1	10.6
1.6 - 1.7	5.7	12.7
1.8 - 1.9	3.3	9.5
1.9	0	-

2. Vitality, respectively daily gain

Because vitality can best be expressed in a good daily gain, this parameter can be used to measure the influence or effect of farrowing accomodation on the well being of the piglet. Taking into consideration the total period from birth to weaning, the farrowing accomodation did not affect daily gain significantly. This daily gain amounted respectively to 171 g $\pm$ 50 g with and to 169 g $\pm$ 49 without farrowing box. By correlating daily gain to birthweight and farrowing accomodation, the following results are noted on table 3.

Table 3 shows that the difference in response to the two farrowing systems under investigations are small or neglectable.

TABLE 3   Daily gain as related to birthweight and farrowing accomodation

Birth-weight (kg)	With farrowing box		Without farrowing box	
	n	Daily gain (g)	n	Daily gain (g)
0.8	6	111	3	150
0.8 - 0.9	14	145	16	137
1.0 - 1.1	38	153	37	147
1.2 - 1.3	88	161	72	158
1.4 - 1.5	88	169	84	178
1.6 - 1.7	84	186	69	183
1.8 - 1.9	9	209	19	198
1.9	2	212	-	-
Average		171		169

188

CONCLUSION

In order to improve well being of the piglets in the period immediately after birth a farrowing pen was adapted with a farrowing box. The farrowing box however did not seem to improve the well being of the piglets as measured by mortality rate and vitality, expressed as daily gain.

REFERENCES

Aumaître, A. 1977. Des maternités bien aménagées reduisent la mortalité des procelets. L'élevage porcin, n° 67,, pp. 39-43.

Braude, R. 1972. The potential for improving sow productivity with practical reference to early weaning. In "The improvement of sow productivity" (Ed. A.S. Jones, V.R. Fowler, J.C.R. Yeats). (The Rowett Research Institute, Aberdeen) pp. 3-13.

Dreihsig, K. 1978. Erhöhung der Aufzuchtergebnisse bei Ferkeln durch zwechtmässige Umweltgestalting im Abferkelstall. Tierzucht, 32, 330-334.

English, P.R., Smith, W.J. and MacLean, A. 1977. The sow-improving her efficiency. (Farming Press, Ipswich Suffolk).

Hughes, P.E. and Varley, M.A. 1980. Reproduction in the pig. (Butterworths, London).

Robertson, A.M. and McCartney, A. 1980. Piglet farrowing boxes. Farm Building Progress, 59, pp. 15-16.

PRODUCTION PERFORMANCE IN LAYING HENS KEPT UNDER DIFFERENT HOUSING CONDITIONS

Rose-Marie Wegner

Institut für Kleintierzucht
der Bundesforschungsanstalt für Landwirtschaft
-Braunschweig-Völkenrode-
D 3100 Celle, Dörnbergstr.25-27, F.R.G.

ABSTRACT

Since 1977 investigations are carried through with laying hens in different housing systems. Comparisons over 3 laying years between hens in deep litter with or without range or in cages resulted in higher egg number per hen housed and egg weight in cages, lower feed consumption, mortality and body weight in cages, less dirty eggs and bacterial contamination with eggs produced in cages, better shell quality and yolk colour of eggs produced in cages. -

Experiments with the aviary system showed good production performance. Disadvantages were: higher percentage of broken and dirty eggs in two types of nests, dirty feathers, bad feathering caused by feather pecking.

Trials with the get-away cage system, testing different bird densities per cage and four different hybrids in comparison to the conventional cage system resulted in a tendency to a higher percentage of dirty and cracked eggs produced in get-away cages. There also was an influence of the kind of hybrid - light hybrids had better performance in conventional cages.

INTRODUCTION

Within the last six years investigations with laying hens started at our Institute in Celle measuring production traits as well as behaviour with regard to 3 main subjects:

1) comparisons of behaviour and production performance between the different conventional housing systems deep litter with range, deep litter and cage battery,

2) possible changes and improvements of the conventional deep litter system by using a kind of aviary system and

3) possible changes and improvements of the conventional cage battery system by using a kind of get-away cage.

In all of these experiments there are considered not only production performance and behaviour, but also economics and hygienic aspects concerning birds as well as birds' products (eggs).

Within this paper a short report is given about the main results of production performances in our first experiments with laying hens in the before mentioned different housing systems to point out, whether there might be any indications with regard to bird's welfare.

1) INVESTIGATIONS CONCERNING THE COMPARISON OF THE CONVENTIONAL SYSTEMS

From 1977 - 1980 we were carrying through three experiments, each over a 52 weeks laying period. We used laying hens of the same light hybrid in the first two trials. In the 3rd experiment half of the same light hybrids and half of medium heavy hybrids were used. At the age of 20 weeks the laying pullets were distributed into the three different housing systems - deep litter with range, deep litter and cage battery. In each trial there were 2304 hens at the beginning, that means 768 birds per housing system divided into 4 pens with 192 birds per pen. The performance results are shown in tables 1 and 2.

Summarizing the main results, there have been found differences in performance between the conventional housing systems as follows:

Egg number per hen housed and egg weight - higher in cages
Feed consumption, mortality and body weight - lower in cages
Shell quality and yolk colour - better in cages
Dirty eggs - and bacterial contamination - less in cages. Most of these traits in which we found differences between the systems are related to economical or hygienic aspects. The difference in mortality - higher mortality in the deep litter system with or without range - caused by wild birds or by cannibalism or partly by internal parasites may be related to bird's welfare.

We hope to be able to continue these experiments with other hybrids during the next years.

2) INVESTIGATIONS CONCERNING CHANGES IN THE DEEP LITTER SYSTEM

We are investigating several types of the so-called aviary system, characterised by an increase in bird density to 10 or 15

TABLE 1 Production Performance of Laying Hens in Different
Conventional Housing Systems

	Deep litter with range	Deep litter	Cage battery
A) Light hybrid (LSL) 3 years, 52 weeks per laying year *)			
Egg number / hen-day	291	286	287
Egg number / hen-housed	266	267	279
Egg weight g	60.8	60.6	61.1
Eggs/ hen/day g	48.6	47.6	48.2
Eggs/ hen/year kg	17.7	17.3	17.6
Feed/ hen/day g	134	132	123
Feed/ hen/year kg	48.8	48.0	44.8
kg feed/kg eggs	2.75	2.78	2.56
Body weight 72 weeks kg	1.96	1.90	1.87
Mortality %	16 **)	10 ***)	6
B) Medium hybrid (Warren), 1 year, 52 weeks per laying year *)			
Egg number / hen-day	290	262	278
Egg number / hen-housed	262	249	273
Egg weight g	62.8	63.3	62.8
Eggs/ hen/day g	50.0	45.5	47.9
Eggs/ hen/year kg	18.2	16.6	17.4
Feed/ hen/day g	138	138	126
Feed/ hen/year kg	50.1	50.1	45.8
kg feed/kg eggs	2.76	3.03	2.63
Body weight 72 weeks kg	2.41	2.27	2.43
Mortality %	22 **)	13 ***)	4

*)   21-72 weeks of age of the hens
**)   mainly caused by wild birds
***)   mainly caused by cannibalism

hens per m² floor area by using perches or small platforms with
feed and water supply in different high levels between the floor
and the ceiling of the house. We were beginning these experiments in April 1978. Until to-day 3 trials are finished, the
fourth experiment is still going on, the fifth is planned for
1983-84. We tested this aviary system with light hybrids (LSL

or Shaver) in two different houses: a conventional house and in a low-cost half-round house covered with polyethylen foil. Table 3 shows the basic data of these experiments.

TABLE 2   Egg quality of Laying Hens in Different Contenvional Systems

	Deep litter with range	Deep litter	Cage battery
A. Light Hybrid			
Deformation μm	44	44	42
Shell percentage %	9.4	9.4	9.6
Albumen height mm	6.3	6.3	6.4
Yolk colour	13.7	13.8	14.2
Cracked eggs %	7	7	6
Dirty eggs %	9	12	2
B. Medium hybrid			
Deformation μm	46	46	45
Shell percentage %	9.2	9.2	9.4
Albumen hight mm	6.1	6.0	5.9
Yolk colour	14.0	14.3	14.3
Cracked eggs %	4	4	5
Dirty eggs %	3	8	1

The results of the first 3 experiments are shown in table 4. The number of eggs collected was nearly normal in experiment 1 and 3, it was lower in experiment 2, probably caused by too many eggs broken and eaten by the hens. The percentage of dirty eggs was too high in all three experiments caused by too many eggs broken by the hens in the large size litter nests. In the roll-away nests the nest floor was not satisfactory - too many cracked and dirty eggs were produced. Feed consumption was very high in experiment 3 because of heavy feather-pecking and there- fore very bad feathering of the hens which already started in the 3rd month of the laying period. Nearly half or two-third of the body of many hens were without any feathers. Only 2% of the hens showed a good feather score at the and of the trial.

TABLE 3   Basic Data of the Experiments with the Aviary System

Experiment No.	1	2	3	4	5
Date	Apr.78/79	Feb.80/81	Oct.81/82	Aug.82-March 83	Feb.83/84
Laying period weeks	52	52	52	36	52
Type of house	low-cost	low-cost	low-cost conventl.	low cost	conventl.
Number of houses/pens	1	2	2	1	3
Number of hens/house	1000	1000/1500	1500/1500	1500	600/1500/1500
Hens/m² floor	10	10/15	15/15	15	7/15/15
Hybrid	LSL	Shaver	LSL	LSL	LSL
Type of nest	litter	litter	litter/roll-away	litter	roll-away

TABLE 4   Production Performance of Laying Hens in the Aviary System

Experiment No.	1	2		3	
House type	low-cost	low-cost	low-cost	low-cost	conventl.
Bird density/m² floor	10	10	15	15	15
Egg number/hen-day	272	252	254	275	278
Egg weight %	61.0	58.6	58.7	61.9	62.5
Eggs not in nests %	2.0	2.9	3.3	0.7	3.0
Cracked eggs %	1.3	5.0	4.5	6.1	10.5
Dirty eggs %	18	12	14	22	11
Feed/hen/day g	127	121	116	141	140
Mortality %	6.4	6.5	6.0	11.5	9.4
Losses by cannibalism %	1.3	1.2	0.9	2.3	0.3

Mortality was normal in the first and second experiment but somewhat higher in the 3rd trial, partly caused by a higher percentage of cannibalism and diseases of the reproductive system.

In further experiments we shall try to solve the main problems: improve the nest system to decrease the percentage of dirty and cracked eggs, lower the percentage of hens with dirty feathers and searching for possibilities to avoid feather pecking. The aviary system is also investigated in Switzerland, the Netherlands and the United Kingdom.

3) INVESTIGATIONS CONCERNING CHANGES IN THE CAGE BATTERY SYSTEM

The experiments with the get-away cage system we began in 1978. Until now 3 experiments have been finished, the fourth experiment started 3 months ago. In the first experiment we were testing different bird densities in a high and low get-away cage, 80 or 65 cm high, with 4 or 3 perches, with 4 nests and with 1 m² floor area per cage. Bird densities were 20, 25 or 30 birds per high cage and 15, 20, 25 birds per low cage in comparison with 3 or 4 birds in a conventional cage (640 or 480 cm²/bird). The original used dustbath was closed after the first few weeks because too many eggs were laid into it. In the 2nd experiment we were testing different kinds of nest floors to make the nests more attractive for the hens and to avoid eggs to be laid into the dustbath. But it was not possible yet to solve this special problem. In the 3rd experiment we used 4 different hybrids, again in get-away cages without dustbath. The hens, two light and two medium heavy hybrids (LSL, Shaver, Warren, Lohmann brown) were reared on litter or in cages from day-old to 20 weeks of age. The results of the 1st and 3rd experiment are shown in table 5.

In both experiments there was the tendency to a lower egg number in the get-away cages, caused by a higher percentage of eaten and broken eggs. There was also a slight tendency to a higher percentage of dirty eggs in the get-away cage system and to a higher feed consumption in experiment 3 in the conventional cages. There was also an influence of the kind of hybrid on egg performance in different cages. Light hybrids showed a better performance in the conventional cages. The rearing method also influenced performance data: Hens reared in cages had a showed the tendency to higher egg production, lower feed consumption· and partly lower mortaly than hens reared on litter.

In the 4th experiment we are again using the same 4 different hybrids testing another nest floor to reduce the percentage of dirty and cracked eggs.

Investigations of the get-away-cage system are also going

TABLE 5   Production Performance of Laying Hens in the Get-
           away-cage-system

1st experiment 1978/79: Different numbers of birds/cage, LSL

3rd experiment 1981/82: 4 different hybrids reared on litter
                        or in cages

	Get-away-cage high	Get-away-cage low	conventl. cage
1st Experiment			
Birds per cage	20/25/30	15/20/25	3/4
Eggs/hen-day %	78	78	81
Egg weight g	60.0	59.8	60.1
Dirty eggs %	2.6	2.3	1.4
Cracked eggs %	14	12	5
Feed per bird per day g	121	121	121
kg feed per kg eggs	2.60	2.59	2.50
Mortality %	7.1	4.8	4.4
3rd experiment			
Birds per cage	25	20	4
Eggs/hen-day %	76	76	77
Egg weight g	60.8	60.8	61.3
Dirty eggs %	2.3	2.3	1.0
Cracked eggs %	5	5	5
Feed per bird per day g	121	123	124
kg feed per kg eggs	2.63	2.65	2.61
Mortality %	6.3	5.3	5.6

on the Netherlands.

From our experiments with different housing systems for
laying hens we can conclude that differences in performance
traits between the housing systems are mainly of importance
with regard to economical and hygienic aspects. Differences in
mortality, mainly caused by cannibalism or parasites in the
intestine may be related to bird welfare as well as bad feathering following heavy feather recking.

These disadvantages have been observed mainly in the litter systems - conventional or aviary systems, but not or dis-
tinctly less in the conventional cage or get-away cage systems.

We shall try in further investigations whether and under
what conditions it will be possible to overcome these disadvantages.

REFERENCES

Hill, J.A., 1981
    The aviary system. 1st European Symposium on Poultry Wel-
    fare, Køge, Denmark, Proc. 155-123.
Institut für Kleintierzucht Celle, 1982.
    Qualitative and quantitative investigations on the beha-
    viour, performance and physiological-anatomical status of
    laying hens under different systems of housing (free range,
    deep litter and battery cages).
    Research Report 76 BA 54, 655 pages in German, 27 pages
    summaries in English.
Rauch, H.-W., 1981.
    Leistungsergebnisse bei der Legehennenhaltung im Volieren-
    system. Landbauforschung Völkenrode SH 60; 153-157.
Rauch, H.-W., Torges, H.-G. and Wegner, R.-M., 1980.
    Zur Bodenhaltung von Legehennen in Folienställen bei er-
    höhter Besatzdichte (Volieren-System). 6.Europ.Poultry
    Conference, Hamburg, Proc. IV, 99-106.
Wegner, R.-M. and Rauch, H.-W., 1981.
    Zur Haltung von Legehennen im Volierensystem in Folien-
    ställen. Aktuelle Arbeiten zur artgemäßen Tierhaltung.
    KTBL-Schrift 264; 235-244
Wegner, R.-M., Rauch, H.-W. and Torges, H.-G., 1981.
    Choise of production systems for egg layers. The Get-away
    Cage. 1st European Symposium on Poultry Welfare, Køge,
    Danmark, Proc. 141-148.

DISCUSSION

Chairman: J. Langholz (FRG)

Discussion of animal welfare in recent years has tended too much towards complying with the newly adopted animal protection laws and their measures of execution. The function of animal welfare research aimed towards defining minimum requirements for legal control should be considered to be of a subordinate nature. Rather, primary research strategy should focus on animal welfare as one component of animal production systems aimed towards optimizing the animal's biological efficiency by improving its welfare within a given or modified production environment. Thus, the question should not be to identify production traits as indicators of animal welfare but, rather, to identify those parameters of welfare that can provide the basis for ensuring performance stability at a high production level and intensity.

It is therefore quite logical to question the validity of using mortality rate during farrowing as a welfare parameter, as was done by H.Bekaert, Belgium in his evaluation of special piglet boxes. It was agreed that incidence of perishing or mortality rate cannot be more than an indicator of the overall welfare situation and used as a preliminary "rough" evaluation of the farrowing system tested. The interest in reducing losses in piglet production caused by the sow lying on the piglets by providing protected micro-environments within the farrowing systems was evident from the number of questions about details and comments. This may be taken as a sign that drastic changes in the cost structure (e.g. drastic increase in energy costs, capital costs, etc.) will require simpler farrowing systems to be established by examining more closely the behaviour pattern and needs of the newborn animal. During the discussion it became quite clear that the vicinity of the newborn to the dam's udder is of significant importance in this context.

Furthermore, parameters of growth performance and feed supply have to be envisaged in a more discriminative manner when considering their relation to animal welfare. For example, breeding heifers having maximum feed supply and maximum growth, show reduced fertility, which cannot be in concordance with maximum welfare. Apparently, maximum welfare relates to certain optima in feed supply and growth intensity depending on species, sex and developmental stage. Furthermore, it was pointed out that health condition and behaviour are likely to outrule growth performance as a parameter of welfare. Here again, one has to emphasize that attention has to be drawn to the impact of welfare parameters on tissue growth, both in quality and quantity, and not vice versa. Concerning the reference traits, much more attention should be paid to the variability of growth instead of average growth as a reaction to different welfare situations.

Finally, M.Hagelsø's (Denmark) studies on the relation between behaviour and growth performance in pigs underline the importance of feeding intensity for the stage of welfare indicating a clear interaction in growth between feeding intensity and sex (gilts, barrows) due to different temperaments. These studies also reveal that G. van Putten's (Netherl.) hypothesis of tail biting as a distinct parameter of aggression cannot be maintained. There was clear evidence that tail biting also is released by a playing instinct. Apparently, here too, one has to move into multifactorial theorems of behaviour interpretation considering a range of releasing factors due to different inherent reactions and different environmental stressors.

SESSION III

INTEGRATED SYSTEMS OF INDICATORS
RELEVANT TO ANIMAL WELFARE

Chairman: G. van Putten

ADVANTAGES AND PROBLEMS OF USING INTEGRATED SYSTEMS
OF INDICATORS AS COMPARED TO SINGLE TRAITS

D. Smidt

Institut für Tierzucht und Tierverhalten der FAL, Mariensee
3057 Neustadt 1, Federal Republic of Germany

ABSTRACT

Mainly 4 groups of biological indicators are employed in the assessment of animal welfare within animal husbandry and management systems: (1) Physiological, biochemical and biophysical indicators, (2) ethological indicators, (3) pathological indicators including morbidity and mortality rates, (4) production performance. All categories of indicators relevant to animal welfare contain problems either in gaining the information necessary for judgement and/or in interpreting observations or values of animal reactions to environmental situations and/or concerning the present knowledge on correlations between potential indicators and their relevance to animal suffering or well-being. Therefore, the combination of single criteria in an integrated system of indicators improve the basis for judgements relevant to animal welfare in husbandry-, housing- and production systems. Problems are still remaining concerning the composition of integrated systems and the establishment of weighing factors for the criteria included.

The discussions during this meeting have confirmed that the single categories of biological indicators have their special significance as well as particular problems when used in assessing animal welfare.

This seminar has been dealing with 4 groups of biological indicators relevant to animal welfare:

1. Physiological, biochemical and biophysical indicators

2. Ethological indicators

3. Pathological indicators including morbidity and mortality rates

4. Production performance

Their advantages and problems can be summarized as follows:

1) If one looks at the advantages and problems of using <u>physiological, biochemical and biophysical criteria</u> as

indicators relevant to animal welfare, the following 3
statements can be made, concerning advantages:

- These criteria represent sensitive and partly specific
  reactions to acute and chronic environmental stress

- Reliable analytic methods for a number of parameters are
  available

- Physiological values may be helpful for interpreting
  ethological and pathological reactions of animals
  (Andreae et al., 1981a).

    Despite these advantages, however, a number of
problems are remaining:

- Single physiological values are rarely significant, because of various biological rhythms and numerous internal
  and external factors of influence.

- Therefore, the employment of these indicators are usually
  dependent on series of values, which allow for setting up
  profiles. However, constant or temporary recording of
  physiological criteria for the establishment of profiles
  requires large work force and financial investment.

- Finally, further knowledge is required on the correlations between physiological, biochemical and biophysical
  traits and environmental effects.

2) Regarding ethological indicators relevant to animal welfare,
   there is little or no doubt about the significance of
   various kinds of disturbed behaviour. There is still a
   certain need,however, for the exact definition of disturbed
   behaviour and of the significance of its frequency of incidences and of the degree and duration of deviations from
   the "normal" behaviour with respect to animal welfare.
       Many more problems are caused by the interpretation of
   reduced fulfilment of behavioural requirements, which has
   been referred to by Wiepkema (1980) as "Istwert-Sollwert-
   Comparison".

There is virtually no animal housing system without
any reduction in meeting behavioural requirements of the
animal. Therefore the nature, frequency and extent of such
reductions have to be regarded as well as the changes as
such. Moreover, possibilities of altering behavioural requirements by means of selection, adaptation and conditioning, have to be kept in mind (Fewson, 1981; Andreae et al.,
1981b, Zeeb & Beilharz, 1980).

3) The employment of pathological criteria as animal welfare
indicators allows for the following statements:
Traumatic lesions are very significant indicators of animal
suffering, if the judgement is based on frequency, location
and severeness of injuries. Disease levels other than
technopathies may be, but must not be the consequences of
production schemes or of husbandry and housing concepts.
Judgement has to include etiology and epidemiology. Morbidity and mortality rates within herds reflect the quality
of management, but are not necessarily subject to legal
consideration of animal welfare.

4) <u>Production performance</u> has often been critizised, as far as
its suitability as animal welfare indicator is concerned.
First of all, it seems important that such indicators can
only refer to "biological" performance traits, but not to
any kind of economical transformation. The relevance of
performance traits as animal welfare indicators can be put
as follows: Unaffected production performance gives some
indication of animal well-being, but exceptions are possible. Lowered production performance (acute or chronic)
should always be taken into account as incicator relevant
to animal welfare. Extremely high performances can be related to welfare-relevant antagonisms, e.g. stress suscept-
ability in modern meat-type pigs.

From these considerations the following statements can be
drawn: All categories of indicators relevant to animal welfare
contain problems either in gaining the information necessary

for judgement and/or in interpreting observations or values
of animal reactions to environmental situations and/or concerning the present knowledge on correlations between potential indicators and their relevance to animal suffering or
well-being. Therefore, the combination of single criteria in
an integrated system of indicators improves the basis for
judgements relevant to animal welfare in husbandry-, housing-
and production-systems (Adler, 1977; Dawkins, 1980, Grommers,
1980; Smidt et al., 1980; Kämmer, 1981, Rist, 1981). The
problem is, however, how to compose such systems. This can,
at present, probably only be done according to different purposes.

For the practical judgement of housing concepts and
equipment one has probably to concentrate on a rather coarse
combination of indicators. The questions to be answered within such a system would be:

1) What are the incidence and severeness of traumatic
   lesions?

2) What are the morbidity and mortality rates? How are
   symptoms and etiology related to animal welfare problems?

3) Are there any acute or chronic breakdowns in production
   performance?

4) Are there any indications of severely and constantly
   disturbed behaviour or of severe reduction in the fulfilment of behavioural requirements?

5) Physiological traits can at present probably only be employed to improve the judgement if the 4 previous questions
   cannot provide sufficient informations.

Combining all categories of indicators, one should be
able to differentiate between + and - situations, as far as
relevance to animal welfare is concerned. However, questionable findings remain. In this sense, Table 1 represents a
possible system of animal welfare indicators (Smidt, 1981).

Still unsolved at present is the problem of weighing the
different traits included in such a system. It will be necessa-

TABLE 1   Biological indicators with relevance to animal welfare

Categories	Criteria	Grading	Animal welfare relevancy of systems
1. Physiological, biochemical and biophysical indicators	unchanged parameters	+	none
	relatively minor changes pointing to stress	- ?	
	pathological values	-	
2. Pathological indicators including morbidity and mortality	low morbidity and mortality, no pathological data	+	the various combinations differentiately evaluated
	elevated morbidity and mortality, higher incidence of pathological data	-	
3. Ethological indicators	no changes in behaviour	+	
	reduced coverage of ethological needs	- ?	
	anomalous behaviour	-	
4. Production performance	unimpaired production performance	+	present
	acute depression of performance	- ?	
	chronic depression of performance	-	

ry at first to set limitations of tolerance for each trait
according to nature, frequency, duration and relevance to
animal welfare.

If this limitation is exceeded in at least one of the
traits, animal welfare relevance of the management system
considered, has to be regarded as proved.

Moreover, the various categories of indicators should be
weighed. It is too early now to give precise figures for
weighing factors.

If the criteria dealt with had to be ranged according to
their relevance, a personal proposal would be:

1) Traumatic lesions and severe technopathies

2) High morbidity and mortality rates caused by welfare-
   relevant circumstances

3) High incidences of disturbed behaviour

4) Severe reduction in the fulfilment of behavioural
   requirements

5) Physiological indications of stress

6) Breakdowns in production performance.

Further work, however, is necessary to improve integrated
systems of welfare-relevant indicators beyond the state in
which they are applied already now.

REFERENCES

Adler, H.C. 1977: Wohlergehen landwirtschaftlicher Nutztiere
      im Rahmen ihrer Produktion. 28. EVT-Jahrestagung in
      Brüssel, 22.-25.8.77, M/3.02/Z.
Andreae, U., Schlichting, M., Thielscher H.-H., Unshelm, J.
      und Smidt, D. 1981a. Zusammenhänge zwischen dem Verhalten
      und den physiologischen Parametern landwirtschaftlicher
      Nutztiere. Internaltionale Konferenz für angewandte Etholo-
      gie, Gödöllö, 24.-27.8.1981, Kommunikationen, pp. 3-18.
Andreae, U., Pougin, M., Unshelm, J. and Smidt, D. 1981b: Zur
      Anpassung von Jungrindern an die Spaltenbodenhaltung aus
      ethologischer Sicht. KTBL-Schrift No. 281, Darmstadt,
      pp. 32-45.
Dawkins, M.S. 1980. Animal Suffering - The Science of Animal
      Welfare. Chapman and Hall, London, New York.
Fewson, D. 1981: Anpassung im Sinne von gezielter züchteri-
      scher Selektion. KTBL-Schrift, No. 281, Darmstadt, pp.
      155-157.
Kämmer, P. 1981. Indikatoren für Tiergerechtheit von Hal-
      tungssystemen für Rindvieh. KTBL-Schrift No. 281, Darm-
      stadt, pp. 129-140.

Rist, M. 1981: Möglichkeiten und Grenzen der gegenseitigen
     Anpassung von Nutztieren und Haltungssystemen.
     KTBL-Schrift, No. 281, Darmstadt, pp. 158-167.
Smidt, D. 1981. Kriterien für die tierschutzbezogene Beurtei-
     lung von Haltungssystemen. Internationale Konferenz für
     angewandte Ethologie, Gödöllö, 24.-27.8.1981, Kommuni-
     kationen, pp. 109-114.
Smidt, D., Andreae, U., Unshelm, J. 1980. Ist "Wohlbefinden"
     meßbar? - Anmerkungen zu einem Tierschutzproblem - .
     Der Tierzüchter 8, 338-339.
Wiepkema, P.R. 1980. Ein biologisches Modell von Verhaltens-
     systemen. KTBL-Schrift, No. 264, Darmstadt, pp. 15-23.
Zeeb, K. and Beilharz, R.G. 1980. Angewandte Ethologie und
     artgemäße Tierhaltung. Tierärztl. Umschau 35, 603-610.

ASSESSING OF HOUSING SYSTEMS BY COMBINED INDICATORS

R.G. Buré
Institute of Agricultural Engineering
P.O. Box 43
6700 AA Wageningen, Netherlands

ABSTRACT

A high frequency of redirected behavioural patterns in an animal housing system is an important datum concerning animal well-being. Big differences between housing systems can give information about the functioning of an animal in relation to the environment. Other criteria can however be of great value too. The rhythm of activity is also strongly influenced by the environment in which the animals live. Comparing housings systems of weaned piglets differences can be found, deviations from a biphasical rhythm. At the same time a high correlation is found between differences in the rhythm and differences in frequency of redirected behaviour.

INTRODUCTION

There are several possibilities in research on the functioning of an animal in relation to the environment. Ethology, physiology and medical science are disciplines of value. They can be applied seperately, but in a number of cases a combination can be very useful and necessary.
Not only a combination of different disciplines can be used in research on animal well-being, also within one discipline different entrances are possible, as can be shown in the ethological approach.

REDIRECTED BEHAVIOUR

In comparing housing systems mostly attention is paid to the appearance of abnormal behaviour. When a number of elementary patterns of behaviour is occuring less in one system than in another and abnormal behaviour comes instead, one can say that in the first situation the animal has problems in adapting. A decrease of exploration and an increase of redirected behaviour can be seen in several oversimplified environments. In research on an open barn with straw, compared with a flatdeck battery, this can be seen. In the flatdeck battery can be spoken about an oversimplified environment - no litter and 0.2 m^2 per animal - in relation to the straw barn - straw available and 0.5 m^2 per animal -. In the flatdeck battery motivation for exploring is decreased and redirection, as rooting on other animals, massaging other animals and nibbling at other animals is appearing.

In table 1. the relation between straw barn and flatdeck battery is given, in which the frequency in the strawbarn is supposed to be 100.

TABLE 1   Relative figures of some behavioral patterns in two different housing systems for weaned piglets

	Strawbarn	Flatdeck battery
Rooting	100	26
Redirected actions	100	456

Also in the comparison of bare pens and pens with straw in a barn with partly slatted floor for weaned piglets differences can be found. These preliminary results are give in table 2.

TABLE 2   Relative figures of some behavioral patterns in two different housing situations for weaned piglets

	Partly slatted floor with straw	Partly slatted floor bare
Rooting	100	91
Redirected actions	100	192

Although in the bare situation rooting behaviour is still appearing a lot, redirection can be seen nearly two times more than in the pen with straw.

RHYTHMICS

   There are still other ways to get some information about animal well-being. Biological rhythms play an important role in the adaptation of animals. Changes in the rhythm of activity can be seen as a direct reaction on changes in the environment (Baggerman, 1978). Schwartz (1974) mentions that a stable rhythm can be an indication that the environment is trouble-free, and therefore the rhythm of activity can give information about the well-being of animals. From research of Marx and Schrenk (1982) can be found that the biphasical rhythm of activity, as can be seen with piglets just before weaning, still exists after weaning if the animals have enough room available, but is disappeared in an overcrowded situation.

In our own comparisons with the open barn with straw and the flatdeck battery observations are made the first two weeks after weaning. Clear differences are found. In the strawbarn the biphasical rhythm is still present while in the flatdeck battery no peaks of activity appear. In fig. 1 the rhythm is shown in the number of resting animals.

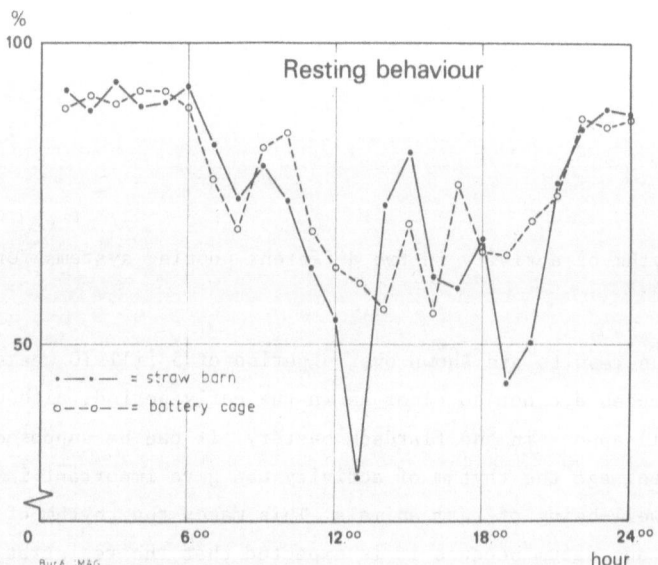

Fig. 1   The number of resting animals in two different housing systems
          for weaned piglets.

When the two approaches are compared and the correlation between differences in the frequency of redirected behaviour and differences in the rhythm
of activity is calculated, high positive correlations appear.

In our present research in strawbarn, flatdeck battery and barn with partly slatted floor it becomes clear that the rhythm of activity can be used
as a criterian for animal well-being, but only with special care. At an
older age the biphasical rhythm becomes less clear, as is shown in fig. 2.

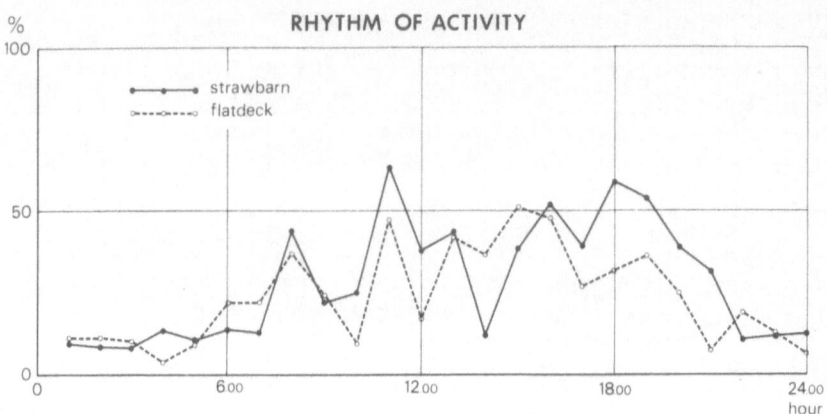

Fig. 2   The rhythm of activity in two different housing systems for weaned piglets.

In this graph the results are shown over a period of 5 till 10 weeks of age. The differences are not so clear as in the early period, although more alterations still appear in the flatdeck battery. It can be supposed that specially at young age the rhythm of activity can give important information about the well-being of farm animals. This makes the rhythm of activity a very important datum because it is expected that the registration can be totally automatized. At this very moment research is going further on this subject.

CONCLUSION

Next to research on the different abnormal behavioral patterns also the studying of the rhythm of activity can give important information a- bout the well-being of farm animals. The influence of the age on the rhythm of activity must however further be understood to make a clear statement about different housing systems.

REFERENCES

Baggerman, B. 1978. Ritmiek. In: Dijkgraaf en Zander, Vergelijkende dier- fysiologie.

Marx, D. and H.-J. Schrenk. 1982. Der Aktivitätsrhythmus von Ferkeln und seine Beeinflussung durch Licht und Futtergabe. Berl. Münch. Tier- artzl. Wschr. 95, 10-16 (1982).

Schwartz, H.J. 1974. Untersuchungen an Mastschweinen zum Problem der quan- titativen Verhaltenserfassung bei landwirtschaftlichen Nutztieren. Diss. Berlin.

APPLICATION OF AN INTEGRATED SYSTEM OF INDICATORS
IN ANIMAL WELFARE RESEARCH

M. C. Schlichting, U. Andreae, H.-H. Thielscher, D. Smidt
Institute of Animal Husbandry and Animal Behaviour Mariensee
Agricultural Research Centre Braunschweig-Völkenrode
Trenthorst, D-2061 Westerau, Federal Republic of Germany

ABSTRACT

The relationship of reactions of farm animals to different
housing conditions is studied in cattle and pigs. Methodical
possibilities are shown by some examples of using several
ethological and physiological parameters. The results will be
interpreted in a relative way in relation to the conditions of
the experiments.

INTRODUCTION

In order to form an opinion about animal welfare in
intensive husbandry it is necessary to use an integrated system
of multiple indicators or parameters. Figure 1 gives a general
view of our working conception of the use of different
paramters by simultaneously recording. The recording of
biological expression of animal welfare is possible in two
ways: first by the recording of behaviour and, secondly, by
collecting biochemical or biophysical data.

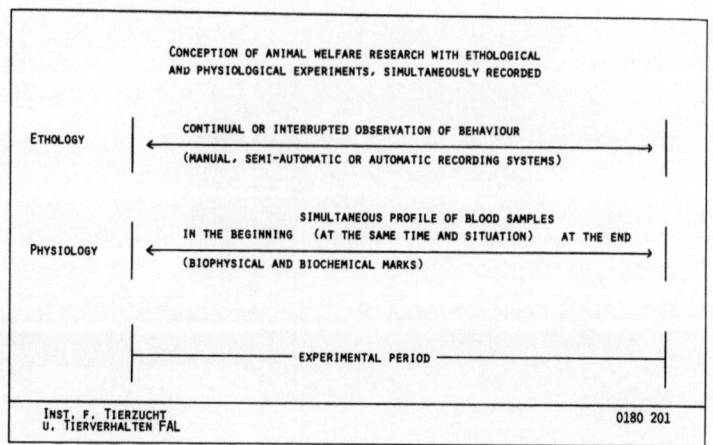

Fig. 1

The combination of these methodical possibilities depends on the experimental question. In using physiological parameters, wie must consider the influence of the blood sampling, a factor which is especially important when we searches for stress-related parameters. For this reason, venipuncture is not used in our experiments, but rather, blood is always collected via catheters implanted into the vena jugularis.

ETHOPHYSIOLOGICAL STUDIES IN CATTLE

Our experimental aim is given by the actual environmental questions of husbandry. It is necessary to prove whether there is a relationship between behavioural and endocrinological changes and environmental conditions. But before we must look what happened in normal situations. So the figure 2 shows for example that the control of adrenaline-concentration depends on different behavioural situations. If there is a relationship of behaviour to endocrinological data, it should be possible to compare endocrinological results obtained under abnormal behavioural conditions with the results obtained from animals showing normal behaviour. Therefore, it is necessary to get blood samples without any disturbing.

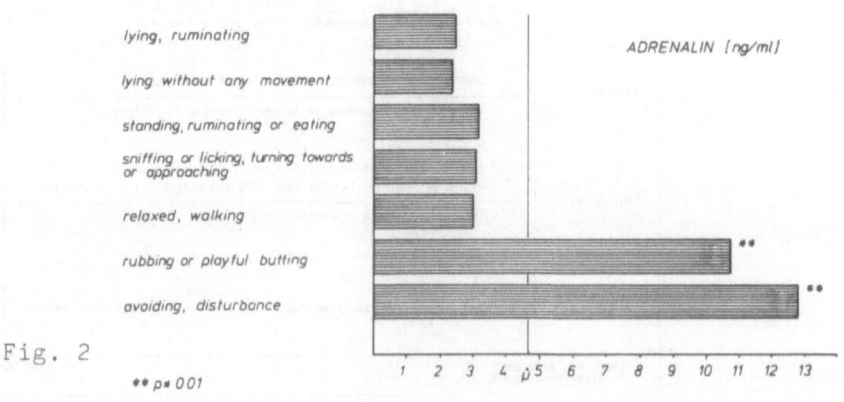

Adrenalin concentration and behaviour in bulls

Fig. 2

** p = 0.01

Figure 3 shows resembling connections under different conditions of husbandry by the cortisol concentration in blood plasma of bulls kept in different loose housing systems during lying down. Lying-down on slatted floor is marked by a higher level of cortisol-concentration and a longer time, as compared to the deep litter.

Cortisol concentration in behaviour of bulls lying down

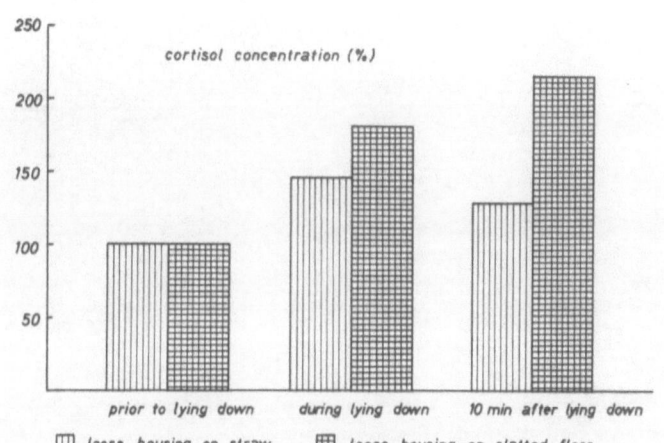

Fig. 3

These short-term-experiments are also applicable to studies of adaptation if blood sampling is necessary (short-term means that it is only a momental situation). During the adaptation phase we observe changes in behaviour as well as in endocrinological results. In this way we set up an development of adaptation of bulls kept on slatted floor.

Integrated experimental methods which analyze a whole phase of adaptation should be more extensive. In a special experiment we test young cattle under different stocking conditions, trying to analyze whether the stocking conditions influence behaviour, health or productional factors.

Figure 4 shows the model of our experiment. Here are shown different stocking conditions of heifers. The following

INTEGRATED SYSTEM OF INDICATORS IN
ANIMAL WELFARE RESEARCH

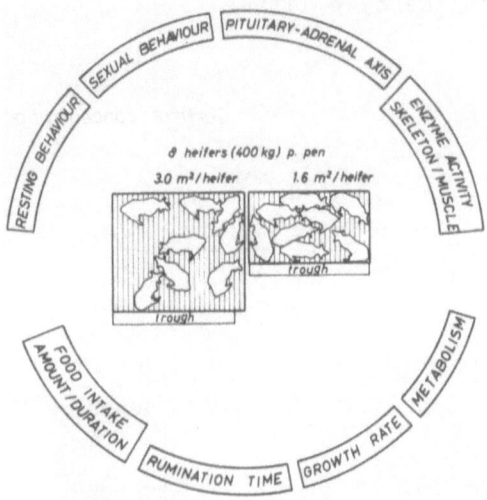

Fig. 4

parameters will be included:
- time of eating, time of ruminating
- number and time of lying phase and the whole time of
  lying
- coefficient of fecundity
- gain in weight
- cortisol concentration in blood plasma
- alkaline phosphatase and creatine kinase as indicators
  of stress to muscle and skeleton.

Preliminary results indicate that there are differences
in behaviour under different rates of stocking. Therefore, it
must be tested whether the rate of stocking causes a load or
stress.

ETHOPHYSIOLOGICAL STUDIES IN PIGS

As part of the animal welfare research conducted in our
Institute, we test influences of different housing systems
based on ethological parameters of gilts as an example of pig
stable problems. Further, we test whether the changes of
environmental situations cause a load or strain on the animals.
This means, for instance, whether there is a difference
between loose housing or individual housing. The change in
housing systems is equal to changes of sow management under
practical conditions. The ethobiological data include
ethological and physiological parameters. The behaviour of the
animals is observed during different phases and blood is
collected according to a special time plan.

Figure 5 explains the system of this experiment. The
animals are subjected to the different types of housing systems
sequentially as well as simultaneously in two models. It is
thus possible to get a vertical as well as horizontal line of
results during the experimental period.

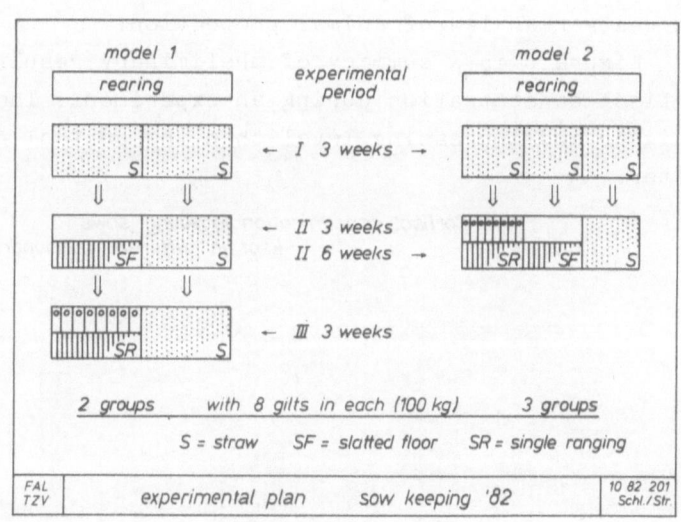

Fig. 5

Table 1 shows preliminary results of behavioural
activities. Special activities are different from one phase to
another, some activities are increased, others have disappeared
because they can not be conducted any more.

Table 1 Frequency of behavioural activities in young sows in different
husbandry systems (n = 2 x 8)

Behaviour	1st week S	2nd week SF	4th week SF	5th week IH	7th week IH
Aggression	4.00	7.19	4.29	0.19	-
Rooting	8.19	9.19	6.69	4.35	4.10
Gnawing at objects	0.5	0.35	0.13	3.92	6.01
Rubbing themselves	2.19	3.66	9.47	3.16	15.5

S = Group housing, straw

SF = Group housing on slatted floor

IH = Individual housing without straw

However, in all this cases, it is not exactly clear,
whether changes of behaviour cause a load or strain on the
animals.

The biochemical parameters are expected to show reactions
related to special organs. It is further assumed that strain
are less than that, what is called pain, suffering or hurt in
the Westgerman law of animal protection.

Figure 6 is a summary of preliminary results and shows the
cortisol concentration during an experiment. The test group has
a higher and increasing level than the control group on deep
litter.

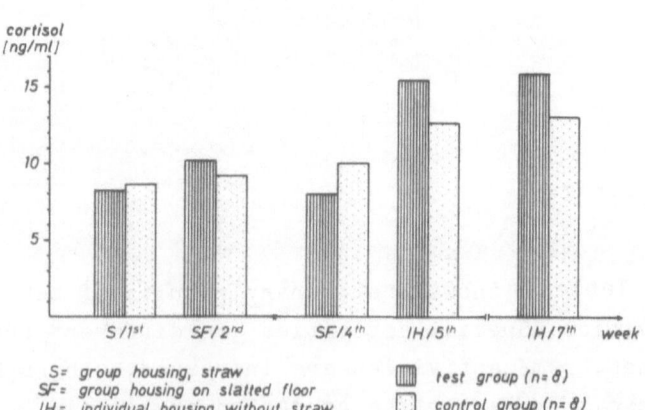

Cortisol concentration in young sows
kept in different husbandry systems (1981)

Fig. 6

CONCLUSIONS

The experiments, which are still being conducted, over methods of multiple indicators relevant to animal welfare, may show the meaning and the necessity of the concept of such research.

In our opinion it is not sufficient to evaluate the complex situation of husbandry with only one parameter. This is particularly important if several questions are attempted to be answered by one experiment and especially if legislative consequences arise. It must not be forgotten, that the results should not be interpreted in an absolute sense. The results must be seen in relation to the conditions of the experiment and to the different systems of husbandry. Comparison of systems of husbandry must be interpreted in a relative way.

REFERENCES

Andreae, U., Thielscher, H.-H., Unshelm, J. and Smidt, D. 1982. Ethological and physiological conditioning of young cattle in intensive housing systems. In "Welfare and Husbandry of Calves", pp. 70-76, CEC-Seminar 1981, published for the EEC by M. Nijhoff, The Hague/Boston/London.
Dantzer, R. and Mormede, P. 1981. Can physiological criteria be used to assess welfare in pigs? In "The Welfare of Pigs", pp. 53-73, CEC-Seminar 1980, published for the EEC by M.Nijhoff, The Hague/Boston/London.
Schlichting, M. C. 1980. Simultane Erhebung multipler tier-schutzrelevanter Indikatoren. In "Tierschutz in der land-wirtschaftlichen Nutztierhaltung", Landbauforschung Völ-kenrode, Sonderheft 53, 38-43.
Schlichting, M. C., Andreae, U., Thielscher, H.-H., Unshelm, J. and Smidt, D. 1981. Biologische, tierschutzrelevante Indi-katoren zur Beurteilung der Tierschutzgerechtigkeit von Haltungssystemen. Züchtungskunde 53 (5), 359-363.
Schlichting, M. C., Stuhec, I., Thielscher, H.-H., Unshelm, J. and Smidt, D. 1981. Ethologische und verhaltensphysiologi-sche Aspekte der Haltung von Jungsauen in unterschiedli-chen Haltungssystemen. Der Tierzüchter 33 (10), 431-433.
Thielscher, H.-H., Schlichting, M. C. and Smidt, D. 1981. Venöse Dauerkatheter bei Schweinen in Gruppenhaltung. Dtsch. tierärztl. Wschr. 88 (1), 33-34.

222

DISCUSSION

Chairman: G. van Putten/Netherlands
     The third session focussed on integrated systems of indicators rele-
vant to animal welfare. To the paper of D. Smidt/FRG a pleade was made
for a more positive approach to health and well-being. Looking for signs
of health provides a different answer than looking for signs of illness.
The behavioural signs of well-being are different from those of a reduced
level of well-being.
     It was answered that the welfare act requires indications of a re-
duced level of well-being. That is, such indicators are legally necessary.
It was further pointed out that positive indicators are difficult to use.
The presence of indicators of well-being is no guarantee for the absence
of other indicators that the needs of the animals are not satisfied.
     In response to the next contribution (R.G. Buré/Netherlands) it was
asked, how the behaviour could be documented as "redirected activity".
It was answered that the determination was based on a comparison to the
situtation in the reference housing system. The use of such a classification, however, was questioned since a behaviour (in this case explorative
behaviour) in the reference system could be regarded as redirection also,
when compared with a more natural environment.
     To a question about how the quality of a substrate should be in order
to be regarded as the adequate stimulus, no clear answer could be given.
     The discussion then focussed on biological rhythms, in this case,
biorhythm of activity. One question was raised, whether the biorhythm was
based upon individual observation and, if so, how it changed with age.
     The biorhythm of activity in the study in question was observed of
the group as a whole and compared with that of the reference system. It
had been observed to change with age, but was also under the influence of
environmental changes. Changes of the biorhythm should therefore always be
correlated with other phenomena.
     It was stated that the synchronized biorhythm is not a valid indica-
tion of well-being. Rhythms of activity are easily adapted to environmental changes and therefore useless as an indicator of affected well-being.
It is just a sign that the animal is able to cope! Furthermore, many influences on the biorhythm may completely have been overlooked.
     It was answered, that in the experiment referred to, the environment
did not change and that, consequently, the biorhythm is a useful indicator.
     A further criticism was stated that absence of biorhythms was a good
indicator of well-being. In the present study, however, only changes in the
rhythms were recorded, not the presence or absence of these rhythms.
     Comments on the paper of M.C. Schlichting/FRG centered first, on
whether the two types of indicators, cortisol levels and behaviour, were
correlated and, if so,whether they supplemented each other. Secondly,
whether differences in cortisol levels were statistically significant. Due
to the preliminary state of the data no concrete answers could be given.
     The general conclusion of the chairman was that agreement was reached
in the necessity for the use of integrated systems, although actual data
were still missing.
     Futhermore, comparison of husbandry systems is a useful approach.
In order to understand results from such experiments, however, better
reference systems should be developed. It was proposed that a special
seminar was arranged focussing on how to design such systems. To this
suggestion it was cautioned against the use of a standard reference system,
since such a system then easily could be considered the ideal system.

SESSION IV

APPLICABILITY OF INDICATORS IN ANIMAL WELFARE RESEARCH
AND IN CONTROL PROCEDURES

Chairman: J.B. Ludvigsen

APPLICABILITY OF INDICATORS IN ANIMAL WELFARE RESEARCH

J. Unshelm

Institut für Tierzucht und Tierhaltung der Universität Kiel
D-2300 Kiel, Bundesrepublik Deutschland

Applicability and possibilities for application of indicators
for assessing welfare can be classified as follows

1. kind of parameters
2. field of reaction to be covered by the used parameters
3. fundamental range of application.

First the kind of parameters. There have been many studies
published about the necessity of interdisciplinary structure
in the area of research as well as practical application. This
is understandable if we consider the number of special fields
the following groups of parameters come from:

1. the behaviour of animals
2. biochemical and other physiological parameters
3. clinical diagnosis
4. patho-anatomical changes
5. percentage of losses

Dependent on the methodical, organisational and financial
possibilities it should be tried to use parameters of all the
fields mentioned above. If this ist not intirely possible it is
also not possible to give sufficient evidence. From a certain
point, perhaps if only a single parameter is registered, the
value of the statement will be reduced so that misinter-
pretations would be unavoidable. This point is a very
important one. The most important problem is the inability
of politicians to distinguish scientific from unscientific
studies. Therefore unwise political decisions base on mis-
interpretations of so called scientific results from in-
sufficient studies. This point is important not only for the
kind but also for the choice of parameters and in the same
manner for the following points which I am going to discuss
here.

The second point in classifying animal welfare indicators is
the field of reactions to be expected. It is important to
differentiate between unspecific stress reactions of animals
kept under certain environmental conditions and disturbed
function of specific organs caused by environment or by high
performance. To explain: housing conditions, but performance
demands too, often cause an unspecific strain of the organism.
I will try to explain later but here it is important to point
out the basic principles. As catchwords I would mention: the
concentration of stress indicating parameters such as the
hormones of the adrenal medulla and the adrenal cortex, such
as reactions of biophysical parameters as heart rate,
respiratory rate and blood pressure, or decrease in performance such as milk yield or weight gain often connected
with a reduced feed efficiency. Along with this but also
independent of it, specific reactions are possible. So over-
loadings of the metabolism are often connected with liver
damage sometimes with negative consequences with respect to
fertility. Specific damages appear often in the area of
locomotion because of specific unfavourable floor conditions.
These reactions both the specific as well the unspecific can
be changed relatively as well as reach dimensions in the
absolute. These dimensions allow for significant conclusions
either about a lack of quality in the environmental conditions
or the complete overloading of the whole organism. To make
this clearer: the metabolism of a high performance cow is
already overstrained by relatively small mistakes in regard
to feeding and management. On the other hand a cow with lower
performance in similar conditions would not react at all.
Similarly, certain pigs and breeds with a high muscle building
capacity react to relatively low environmental conditions with
overstraining of acid-base-status, circulation-system and the
skeleton. They tend to a higher incidence of disease, are more
susceptible to infection and there are altogether greater
losses in rearing, in fattening and in transportation than
with breeds or lines with a lower performance capacity. These
kinds of influencing factors must be considered very

consistently when interpreting experiments and results of this nature to avoid false interpretations.

The third point I would like to discuss is the fundamental range of application. Most of the investigations,consequently employing the factors relevant to animal welfare, were started by specialized research institutes with experimental conditions.It is a decisive factor that all these experimental conditions were very good or optimal in regards to the organization, the setting-up of experiments, the animals being used, the scientists and the technicians, the exactness of the methods used in the experiments, the accuracy of the statistical analysis, and the interpretation. There will always be questions which may only be checked in research institutes whereby the results are to be used in practice.

In contrast to this, in future we will need more animal welfare indicators for analysing the situation in practice. This will be increasingly important depending on the fact that it will not be possible to make an optimal decree of animal welfare in every case. For example we have the difference between a perfect and a less perfect manager. Animals belonging to a poorer manager normally have higher rates of metabolic disturbance, technopathies and increased rates of illness and losses in spite of excellent technical prerequisites. On the other hand, there are many examples supporting the fact that an excellent manager is able to compensate for very inferior quality of technical conditions with the result that the performance of the animals and their welfare are not influenced. It would be absurd to make an assessment of these farms based only on law and decree. It is more important to use the reactions of the animals in the different housing systems as welfare criteria. I am sure that this way is the only right one,even if it means a lot of effort on the part of scientists and legislators. It is also clear to me that this situation moves decisions and responsibilities from the centre to the periphery. In addition to the fundamental reflections I mentioned above I have come to this conclusion on the basis of policies in other

countries, for example Switzerland, which is relatively easy
to overlook, or Sweden, which already has a peripheral control
system.

Independent of the attempt to regulate every possible freedom
of decision through strict laws and decrees or to assess the
situation in every single case by commissions in the periphery
there remains the urgency of a common direction with a
statement regarding what may be demanded of animals. This
concept of what may be demanded of an animal which exists in
the German animal protection act urgently needs more precision. Therefore we must decide if it is to be allowed to
have one animal with an abnormal behaviour among a thousand
"normal" animals on a farm when determining the acceptability
of the housing system with respect to the animal protection
act. Or let us discuss the losses in animal husbandry and
during transportation. Obviously we can never develop a
housing system or a method of transportation without one
animal becoming ill or dying. On the other hand housing
systems with extremely high losses should not be allowed.
These systems conform neither to the needs of the animals
nor to the economic situation. There should be qualified
institutions responsible for making decisions in these cases.
Naturally they must take into consideration the variation
rates which are unavoidable in biology.

After having discussed these more fundamental comments I shall
show some illustrations of factors which may be used as
animal welfare indicators. All of them use the animal reaction as criteria.

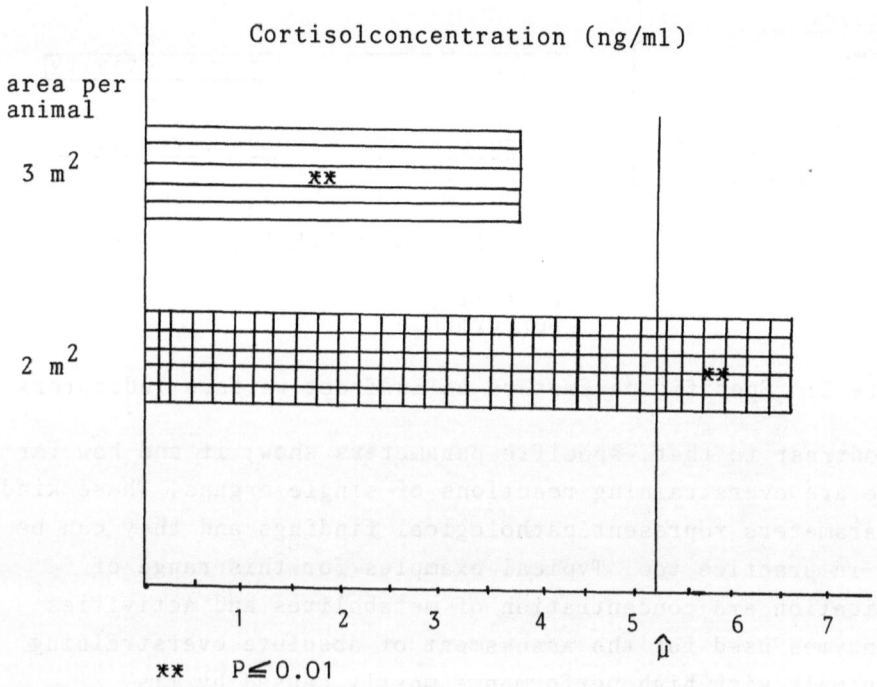

** P≤0.01

Figure 1: Area per animal and cortisol concentration

This figure, a result of my work in this institute (Unshelm, Andreae, Smidt,1982), shows an unspecific reaction. In bulls of the same age hormone profiles were measured using permanent catheters for assessing the reaction caused by differences in the area per animal. As you can see on this figure the cortisol concentration in the bulls, each kept on 2 m^2, were significantly higher than in bulls with 3 m^2 per animal. This unspecific reaction shows only that an area of 2 m^2 per animal is more overstrain than an area of 3 m^2, but it does not show if this strain is acceptable or not. Without a doubt we need further investigations in this field covering a greater area also with respect to overstraining reactions of animals and last but not least the political decision regarding the absolute point of overstraining. It is obvious that this kind of investigation can only be done under institutional conditions and not in practice.

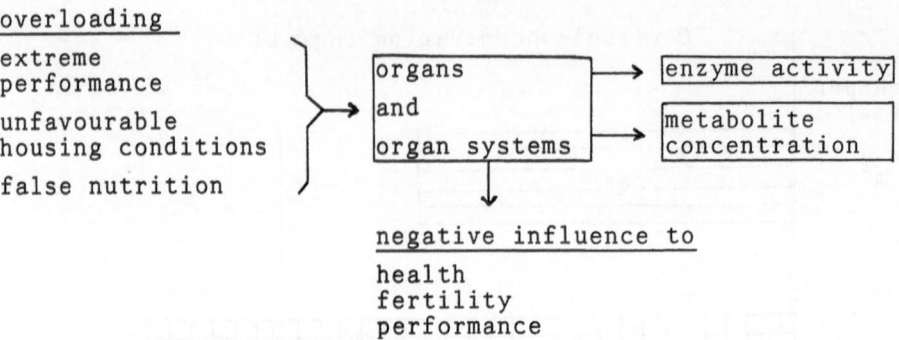

Figure 2:   Specific parameters as indirect welfare indicators

In contrast to that, specific parameters show, if and how far
there are overstraining reactions of single organs. These kind
of parameters represent pathological findings and they can be
used in practice too. Typical examples for this range of
application are concentration of metabolites and activities
of enzymes used for the assessment of absolute overstraining
in animals with high performance mostly caused by mis-
management.

Table 1:   Housing system and losses

housing system	number of farms	losses %	gross margin DM/100 kg
with straw	104	3,07	31,4
without straw	240	3,04	35,5
full slatted floor	28	3,74	29,2

Strukturanalyse der Schweinehaltung in Schleswig-Holstein

This point covers the area of technopathies as discussed. We
have housing conditions with increased losses caused by
environmental and specific organ damages. One example for that
is the leg-weakness in pigs kept on full slatted floors. The
low gross margin for pigs kept under these unfavourable conditions is to be noted.

Table 2: The difference between Schleswig-Holstein and north German unions of community (average 1977 - 1981)

feature	Schleswig-Holstein	north German unions of community
daily weight gain (g)	563	587
feed conversion (1:)	3,47	3,37
losses (%)	3,3	2,6
profits (DM/kg)	2,96	2,95

Strukturanalyse der Schweinehaltung in Schleswig-Holstein

The last point I would like to discuss is the well-known problem of stress-susceptibility which is very different and dependent on performance. So we see negative reactions to housing conditions in meat-type pigs which cause no sort of injury to health and welfare in breeds which are more resistent to stress. As this table shows pigs with high muscularity from Schleswig-Holstein are more unfavourable with regard to daily weight gain, feed conversion and losses than pigs with a slightly lower performance. The profits, however, are similar because of the higher classification of pigs with better muscle-building-capacity.

From my experience with politicians there is nothing which makes a scientist more popular than when he produces a cut-and-dried, foolprove solution. My lengthy involvement in biological relationships has convinced me, however, that this is not the way to be gone by scientists. These things can not be dealt with so easily. Therefore if I have not succeeded in offering a practicable formula for the intricate problems of the animal welfare indicators, I suggest the following: if biology can't be made to fit politics, we should try it the other way around.

REFERENCES

Strukturanalyse der Schweinehaltung in Schleswig-Holstein. Published by Landwirtschaftskammer Schleswig-Holstein, Kiel, 1982. (table 1 and 2).

Unshelm, J., U. Andreae, D. Smidt: Biochemische Parameter im
    Rahmen tierschutz- und nutzungsbezogener Untersuchungen
    beim Rind.

Advances in Veterinary Medicine
Suppl. to Zbl. Vet. Med., No. 35, 220 - 225 (1982)

THE USE OF ETHOLOGICAL WELFARE INDICATORS IN PRACTICAL CONTROL PROCEDURES

K. ZEEB

Tierhygienisches Institut Freiburg

7800 Freiburg i. Br., Federal Republic

of Germany

ABSTRACT

The German (FRG) animal protection law protects the life and well-being of animals and prohibits infliction of pain, suffering and damage to animals. These demands are based on the concept "species specificity" which is difficult to utilize in practise. Specification of requirements of animals and the means of their satisfaction, as well as avoidance of damage to animals seems to be better approach.

Based on this approach a scheme is presented for judging whether an animal's requirements are satisfied. This scheme is explained with the example of feeding of dairy cattle.

Results of a study of the suitability of cubicles for dairy cattle is then discussed from the aspect of avoiding damage to the animals.

The paper concludes with suggestions for training people in the proper application of ethological animal welfare indicators.

INTRODUCTION

The German (FRG) law for animal protection of 1972 enforces in § 1 protection of life and welfare of animals and also the prohibition of inflicting of pain, suffering and damage to animals. § 2 prescribes the safeguarding of species-specific nutrition, care and housing appropriate to the behaviour and satisfaction of the species-specific needs for movement. The official reasoning behind the law of animal protection was as follows: Consideration of evolutional, adaptational and domestical state of the individual, based on the knowledge of natural science in order to safeguard its well-being, that means the undisturbed occurance of the life processes which are species-specific and appropriate for the animal. Kämmer and Tschanz (1982) argued: "When these laws (also the Swiss law of animal protection) were formulated, it was assumed that methods were known in the field of applied ethology by which such

Table 1. Satisfaction of requirements as to satisfaction of needs

OBJECT	BEHAVIOUR	N E E D	
		REQUIREMENT	SENSATION
Watering place	drinking	water	thurst
Food	eating	nutrition	hunger
lleat source	staying	warmth	freezing
Litter	burrowing	activity	boredom
Companion	staying following	proximity of companion	abondonment
verifiable by natural science			not verifiable by natural science

states could be objectively defined, but this has proved to be an error.
On the other hand, it is reasonable to assume that higher vertebrates do
feel sensations like suffering, well-being and pain on the grounds of
their neural equipment and structure. What can applied ethology contribute
if this assumption is accepted but direct methodological evidence of it
cannot be furnished ?"

REQUIREMENTS AND NEEDS

The ideas of Tschanz (1981) and Kämmer and Tschanz (1982) may be
demonstrated briefly as follows. Ethology cannot determine what happens
in the animal, but only its appearance and how it modifies its behaviour
and its environment.
Features of body, behaviour and environment permit statements about
whether housing and management conditions of the animals fit the needs of
the animals and therefore also the animal protection law.

The are four ways of considering animal management and housing: The

- biological   - ethical
- economic   - legal

point of view. Ethology only deals with the biological considerations.
Basic properties of life are, that organisms build up, maintain and
multiply themselves. This means for livestock husbandry as well as for
animal production: Growth, supply and reproduction. But from these arise
requirements which have to be met.

It is behaviour that enables requirements to be met. If requirements
are not met damage arises. And it is again behaviour that enables avoidance
of damage. For building up, maintaining and multiplying of self, objects
and situations are necessary, which enable requirements to be met and
damage to be avoided by behaviour. Table 1 shows requirements (e.g.water),
the appropriate objects (e.g. watering place) and the behaviours (e.g.
drinking) for meeting these requirements. Natural science can make
verifiable statements about these requirements, objects and behaviours.

Quite another thing is the satisfaction of needs. Needs comprise
both requirements and sensations (e.g. water and thirst). Natural science
cannot make verifiable statements about needs. Note: Tschanz and Kämmer,
in the German language very carefully separate satisfaction of requirements
(Bedarfsdeckung) from satisfaction of needs (Bedürfnisbefriedigung).

236

Table 2. Scheme for judging welfare indicators in animal management and housing systems

Ethological indicators		Non-ethological indicators
ITEM meeting the requirement	BEHAVIOUR	FITNESS of facility	
- Quality	- form	- animals functional measurements	- clinical
- Quantity	- frequency	- number of animals	- anatomical-pathological
	- duration	- soc. integration degree	- physiological-biochemical
	- intensity	- climatical point of view	- biophysical
	and their modifications	- hygienic point of view	- performance of production
		- property of material	- performance of reproduction

What methods are available to ethologists for judging the fitness of husbandry conditions for animals ? Behaviour patterns are categorised into functional domains (e.g. excretion behaviour). Features of the environment are categorised according to the technical procedure of the management and the design of housing (e.g. manure disposal). The inter- action between behaviour and environment produces modifications of the animal and its environment as requirements are met and damage is avoided during build up, maintenance and reproduction of the animal. These modifications of the animal and the environment can provide indicators for the adjustment of behaviour and environment to the animals requirements.

INDICATORS OF FITNEES TO THE ANIMALS REQUIREMENTS AND FOR ANIMAL WELFARE

Table 2 shows a scheme for judging welfare indicators in animal management and housing systems. This scheme contrains four groups of parameters. Three of them are ethological indicators :
  - Item meeting the requirement: Quality and quantity.
  - Behaviour: From, frequency, duration, intensity and their modifi-
    cations.
  - Fitness of facility: Animal functional measurements, number of
    animals, social integration degree, climatical point of view,
    hygienic point of view and property of material.
The fourth group contains non ethological indicators:
Clinical, anatomical-pathological, physiological-biochemical, bio-physical, performance of production and performance of reproduction.

EXAMPLE ILLUSTRATING THE SATISFACTION OF THE REQUIREMENT NUTRITION

Using the example of feeding dairy-cattle we may see the usefulness of the scheme presented above. To judge the item meeting the requirement the food must be appropriate to cattle with regard to structure, digesti- bility and nutrition value. With respect to feeding behaviour of cattle form means the species-specific kind of bovine food intake. Frequency relates to the diurnal rhythm of feeding which alternates with other activities such as ruminating and resting. Duration means that the time for food intake must be long enough for the given feeding conditions. That is, for example, about 10 hours per day in camargue cattle and about 8 hours in cubicle stalls (Zeeb and Bammert 1978). Intensity means the strength of the food intake action.

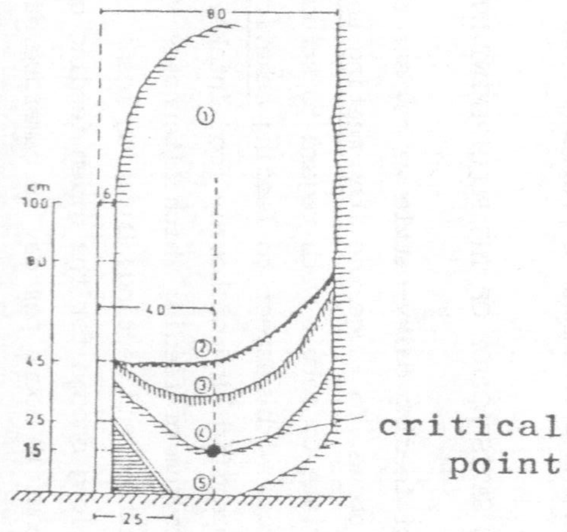

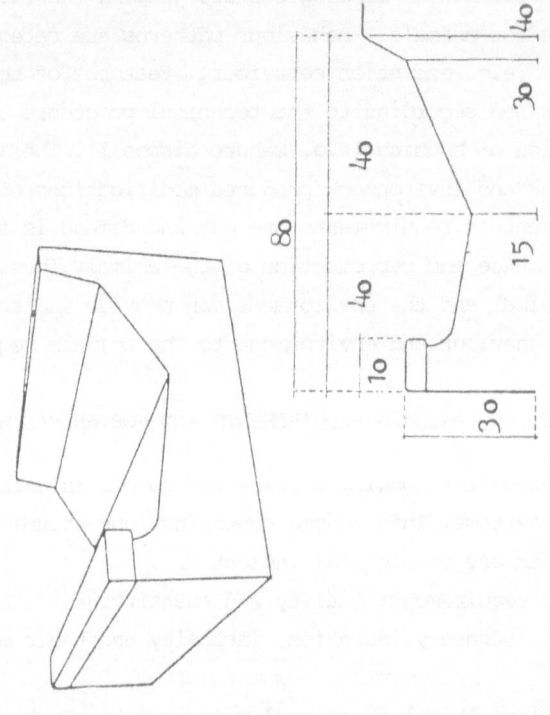

fig. 2. Feeding trough for cows tied in a short-standing facility

fig. 1. Feeding profiles of cattle in a clamp silo

critical point

To judge the _fitness of facility_, the functional measurements of the feeding area must be considered. Fig. 1 shows the profiles which cattle form when feeding from a clamp silo equipped with palisade rack. The depth of the profile, where the animals are able to feed is 80 cm. There is no problem of feeding down to a point of 15 cm from the level of the ground. This point is a function of the size of the animal and the length of neck and head. The last 15 cm of silage cause some difficulty. While cattle graze the spreading of their forelegs and shifting of weight from one to the other allows the mouth to reach the ground easily. It may be acceptable to let freely moving cattle feeding at a clamp silo feed down to the ground periodically for a very short time. In other words this facility fits their needs. That is not the case for cattle tied in a short-standing facility. To fit this facility to their requirements while feeding, the lowest point of the feeding trough must be 15 cm high. The trough must be 80 cm deep. The profile of the trough (fig. 2) should be the same as profile 4 on the silage (fig.1). Concerning the number of cattle at the feeder a feeding place of about 70 cm must be available for each animal particularly when feeding is at set times. When feeding ad lib. a feeding place to animal relationship of 1 to 2 is possible in the case of a socially integrated herd. The palisade rack, into which the cattle must slide by lifting head and neck, prevents serious social interaction between neighbours.

As to climatic factors, feed in the feeding facilities should not be frozen nor spoiled by rain. We now know (Bammert and Zeeb) that feeding activity of cattle is highest during hours with a temperature from 8 to 12° C. From the hygienic point of view, feed must not be contaminated with infected material. The material used for the construction of the facility must not cause injury to the animals.

If, as well, we find no deviation concerning the non ethological indicators such as the clinical etc. we can say that feeding of cattle is fitted to their requirements as the points checked above are fulfilled. In other words: When these conditions are given, cattle are able to meet their food requirements and we can reasonably assume that the development of well-being will be possible.

THE AVOIDANCE OF DAMAGE IN CUBICLES

Kämmer and Tschanz (1982) give good examples for an environmental feature which nowadays often leads to injury in cattle: The cubicle box. During their studies of dairy-cattle in cubicle houses, they found "that in certain types of boxes modifications occured in lying down, lying and getting up. Typical changes in lying down were rare, but connected in most cases with contacts of animals body with parts of the box. For the getting up they found one new movement pattern (horse-like getting up), and also contacts with parts of the box. This was caused by a lack of movement space, which in most cases did not permit species-specific movement patterns to occur without heavy contact. Such contacts often lead to injuries, types of behavioural modifications and their frequency of an animals body contact with parts of the equipement and their consequences can be described by exact methods."

Kämmer (1979) recommends a practical method for the diagnosis of inadequate fit of cubicles to the animal requirements. Such inadequate fitting is given e.g. when:

- On average cows tread more than four times before lying down.
- One or more cows don't lie down during more than two main lying periods.
- More than 10 % of the cows lie outside the cubicles.
- More than 10 % are lying with serious contacts of bony parts with the cubicles side facility.
- Blood and parts of the skin are visible on parts of the cubicles.
- Parts of the cubicles are damaged.

This is only an extract of the criteria Kämmer presented.

The value of 10 % of cows is based on the fact, that the recruitment of a herd during one year is about 10 % of the animals. That means during every inspection some animals are in a state of adaptation to the facilities.

These quantitative values are minimal limits for avoiding damages.

COMPARISON BETWEEN THE FITTING OF DIFFERENT FACILITIES FOR THE SAME PURPOSE

What has been described so far is a useful valuation key for judging the fitting of facilities for the requirements of dairy-cows. Much more difficulties are posed by the comparison between two different kinds of facilities.

Agricultural engineers often favour one technical facility for technical reasons. An example is the cow-trainer. Nowadays in our country Holstein-Friesian cows are frequently tied up in short-standing facilities with a stand of 165 cm in length. To keep a stand of such a length clean the tying system does not permit much movement. When equipped with a cow-trainer the stand length can be longer, e.g. 180 cm, and looser tying becomes possible. The cow-trainers electroshock forces the cow to step backwards before urination or defaecation, because of her hump while urinating or defaecating. The engineer now thinks: The cow learns to avoid the electroshock, gains more movement space and has a better hygienic situation. In my experience cows are unable to completely avoid shocks when they come on heat and while self-grooming. That means that, while it is acceptable to guide an animal's movements by physical structures which cause no pain, use of electro-shock in a tighty restricted situation makes it impossible for the cow to escape pain.

Up to now we have had no possibility of evaluating what is better for the cow: Small movement space or no escape from shock. For me it seems to be one of our most urgent obligations as applied ethologists to find out methods of evaluating the fitness of environmental features and behaviours with respect to the animals requirements.

People involved in ethological judgement of animal production could, in my opinion, be trained as shown so far, in the way of finding out whether animals are capable of satisfying their requirements and avoiding damage in their housing system or not.

SUGGESTIONS FOR TRAINING PEOPLE IN THE PROPER APPLICATION OF ETHOLOGICAL ANIMAL WELFARE INDICATORS

In the sixties some of us started pioneering work to convince livestock husbandry people that applied ethology is a good instrument for avoiding reduced performance of livestock. We did not start by saying: "Animal production is cruel, free these poor animals!" No, we collaborated hand in hand with interested farmers to find out the ethological errors of our livestock production system and how we could apply the methods of basic ethology which Lorenz, Tinbergen and other wise people had developped. These basic ethologists did not translate their methods of studying species in the wild to their domesticated descendants used in animal production. We applied ethologists, most of us coming from the veterinary or agricultural sciences, had to adapt methods ourselves and also to gain the confidence of farmers for this new science. While this occured in the application of ethology to cattle, horses and (minor) pigs generally it was not so with poultry in our contry. We now have the situation that applied ethologists are a kind of friend to the cattle, horse and pig producers, but a kind of - excuse the expression - enemy of poultry producers. Where applied ethologists were seen as friends, livestock producers are adapting their findings. Hence it seems obvious that the above outlined training of people involved in animal production should be on the basis of friendship and cooperation and not on the basis of judgement and revenge.

Applied ethology itself seems to be at present in a state of change. For example, again, much difficulty has been found in applying the description "species-specific" in discussing animal welfare. That is why we are looking for new approaches and methods. This is my reason for drawing your attention to the concepts "satisfaction of requirements" and "avoidance of damage". They are not to be used for substitution of the term "species-specificity" but help for the practical fulfilment of it. And they could help to more readily determine indicators of animal welfare.

Ethological indicators can stand on their own and do not need support from any other science. On the other hand, depression of performance, poor health, etc. are also indicators of impaired welfare. It was not the purpose of this contribution to cover aspects other than ethology.

REFERENCES

Bammert, J. und Zeeb, K. 1983. Der Einfluss meteorologischer Faktoren auf die Aktivität von Rindern. In Vorbereitung.
Kämmer, P. 1979. Untersuchungen zur Tiergerechtheit und ihrer Bestimmung bei Boxenlaufstallhaltung von Milchkühen in der Schweiz. Diss. Uni. Bern.
Kämmer, P. und Tschanz, B. 1982. Animal welfare as judged by ethological methods. Applied animal ethology. 8, 404.
Tschanz, B. 1981. Verhalten, Bedarf und Bedarfsdeckung bei Nutztieren. 13. Int. Arbeitstagung Angewandte Ethologie bei Haustieren, DVG.
Zeeb, K. und Bammert, J. 1978. Effects of climatic factors on cattle under various systems of management. 1. World Congress on Ethology applied to Zootechnics. Madrid.

DISCUSSION

Chairman: J.B.Ludvigsen

The pertinent question of applying welfare indicators to practical animal husbandry and the approach to adjust operating confinement systems and technical devices to existing knowledge of the behavioural needs of confined farm animals is a complicated interplay between the society and producers of food products.

Since the Council of Europe adopted the Convention on the Protection of Animals used for Farming Purposes in 1976, the political involvment in the welfare issue in the EEC has been increasingly important.

At present, however, only limited progress has been made to adjust production to the recommendations of the Convention.

There are many reasons for the delayed application of the Convention to producer levels. Difficulties among the scientists involved in welfare research to reach a common terminology and definitions of animal welfare is quite an important part of the problem, as long as welfare indicator research, still in its youth in a debatory phase, may seem controversial.

Another part of the problem is that some politicians,who are under more or less constant pressure from welfare m vements, animal protection societies and consumer groups, are likely to call for either "black" or "white" statements and decisions from the scientists, although it is generally accepted that the problems involved are neither black nor white.

Nevertheless, the speakers as well as the participants in the discussion agreed on a very important part of the welfare issue. It was repeatedly underlined that education and training of farmers and stockmen is a problem of major importance, as well as several requirements, such as routine inspection of production systems by experts to correct faulty managerial factors and to stimulate the observance of producers and their personal,should attract high priority along with welfare indicator research.

GENERAL CONCLUSIONS

Chairman: J.P. Signoret/France

In each of the fields - physiology, ethology, pathology and production - indicators of animal welfare have been proposed and their scientific value tested and discussed. We will try first to summarize the major
results in each case, their importance for the practical evaluation of
well-being, the questions arising, and the topics requiring further research. In a second part, we will be concerned with the practical use of
such indicators in the assessment of welfare and in the recommendations
to the regulation makers.

1) Existence of indicators, their value, limits and possible improvement

The physiological indicators appear to be correlated to stress, pain and
suffering. Being measures of humoral hormones or neurotransmitters, they
are indicators better used for research purposes than for extensive field
use. However, their significance as indicators of pain or discomfort, or
as physiological reactions leading to adaptation is not thouroughly clear.
Nevertheless, they appear to be usable for the apraisal of the level of
discomfort generated by a given stimulation, condition of living, handling
etc., but mostly under experimental conditions. It seems hardly possible
to use such indicators in the field evaluation on a large scale, due to
the high sophistication of sampling and assay procedures.

Finally, if pain, stress and discomfort were mediated by the same and
single system,it would be possible to interfere with this system by the
use of drugs, which could facilitate the phase of adaptation to a new
situation, and thus to eliminate an important part of what makes such an
adaptation unpleasant.

The ethological indicators, such as disturbed behaviours, stereotypies,
etc. are relatively easily categorized and observed. However, their status
as indicators of the severity of discomfort remains a question. Some behaviours, for inctance, occur in conflict situations and,thus, may help
adaptation. Experiments have shown, that chewing a chain prevents or reduces the adrenal response to frustration in pigs.

The use of disturbed biological rhythms as an indicator is also
questionable: Can we assume that the "normal" activity rhythm is necessary
to welfare and, furthermore, are changes easily accepted unpleasant or relatively indifferent?

A third unanswered question regarding ethological indicators to
stress is whether the display of the total behavioural repertoire is necessary to the welfare of the animal or whether the lack of opportunity to
dustbathe, root, etc. is detrimental to the welfare.

The importance of disturbed behaviours depends on the breed of the
domestic mammal. The possibility for a genetic selection for a better
adaptability to a given environment has to be considered and, in fact, is
a part of the evolution of domesticated animals through the centuries of
animal breeding in captivity.

Finally, a point that has   been treated in this seminar, is that
the indicators of well-being could be very useful in evaluating not only
the negative and detrimental consequences of a given husbandry system, but
the possibilities for favorable ones.

The pathological indicators have an absolute value as indicators of welfare. Mortality, morbidity or presence of disease or wounds cannot be

associated directly with welfare,but their evaluation is a first approach to the validity of a given husbandry system. Increasing health and reducing mortality and morbidity always means increasing welfare.

The production performance has some ambiguous significance: In some cases, any major trouble, physiological or psychological,results in an impairment of production. This is the case with milk or egg production, whereas growth does not seem as sensible. Another ambiguity lies in the necessity to satisfy all the animal's needs. Is ad libitum food necessary for animal welfare? Especially when some level of food limitation improves health on a long term.

The animals selected for a high level of performance are generally more sensitive to any kind of imbalance in management. This is the case in high production dairy cattle, rapidly growing pigs, in broilers, race horses, etc. In such animals, management errors rapidly lead to more difficult situations than in old fashioned management systems. However, management errors cannot be considered to be the rule.

The integration of the various indicators appears to be the only way for a total evaluation of animal welfare. In fact, the various indicators could be relatively independent and eventually, in some cases, could give a different, if not opposite, estimation of welfare. For instance, laying hens on deep litter have a better feathering than laying hens in battery cages but present higher incidences of cannibalism and disease.

2) The use of indicators

As shown in scientific observations and studies, the indicators appear to be objective facts. The problem is to give each of them, or a combination of them, a value for practical apraisal, in order to improve animal welfare.

The various indicators of welfare are obviously of different importance. This suggests the possibility of a hierarchy among them. If this is so, the pathological indicators could be placed at the highest level, being a sine qua non condition of welfare. However, classifying the other indicators is difficult. The place in the hierarchy could be influenced by the opinion of each person.

Thus, a multifactorial approach would probably be more efficient. In considering all the possible indicators, toghether with the others: production, pathology and economic traits, it would be possible to obtain an idea of the welfare status in a given production system. Such an approach appears to be more valuable. Any system represents, in fact, a compromise between differing, and often opposed, technical requirements: for the optimum production, health, economy and welfare. The multifactorial approach is the only one to take into account all the factors contributing to production. Consequently, it is the only one to be really effective in practice.

The use of indicators of welfare for recommendations to regulation makers is, under such conditions, difficult. Rather than focussing on one category of indicators - with the exception of the pathological ones - it would be more efficient to use the multifactorial approach which takes into account all the factors resulting in the production. Any improvement of welfare has to follow such a way to be successful in practical production.

The <u>ethical value of the welfare indicators</u> is critical for their use. It is necessary to bear in mind that the indicators are scientific facts. Provided that animals' health and production is not impaired, the value of the indicators is established from the comparison with ethical references. However, such references are highly variable between European countries, between cities and rural areas, and between socioprofessional categories.

To conclude, welfare indicators - existing or resulting from future research - are objective, scientific facts. Whatever the conditions could be in domestic animals as well as in human beings, life is a compromise that can practically never be maintained at the optimum level in all respects. Welfare indicators will be useful tools in allowing us to improve the living conditions of the animals that man is keeping for his own benefit.

LIST OF PARTICIPANTS

BELGIUM

Dr. H. Bekaert
Rijksstation voor Veevoeding
Scheldeweg 68
9231 Melle-Gontrode

Prof. Dr. J.M. Bienfait
Faculté de Médecine Vétérinaire
Rue des Vétérinaires 45
1070 Brussels (Cureghem)

Dr. Nicks
Faculté de Médecine Vétérinaire
Rue des Vétérinaires 45
1070 Brussels (Cureghem)

DENMARK

Dr. Mette Hagelsø
National Institute of Animal Science
Research in Pigs and Horses
Rolighedsvej 25
1958 Copenhagen 5

Prof. Dr. J.B. Ludvigsen
National Institute of Animal Science
Rolighedsvej 25
1958 Copenhagen 5

Dr. H.B. Simonsen
Royal Veterinary and
Agricultural University
1870 Copenhagen

Federal Republic of Germany

Dr. W. Bessei
Universität Hohenheim
Abt. Kleintierzucht
Emil-Wolff-Str. 46
7000 Stuttgart 70

Dr. Doris Buchenauer
Kölnstr. 141
5205 St. Augustin bei Bonn

Prof. Dr. K. Dämmrich
Institut für Veterinär-Pathologie
d. Freien Universität Berlin
Drosselweg 1-4
1000 Berlin 33

Prof. Dr. A. Grauvogl
Bayerische Landesanstalt für
Tierzucht
Prof. Dürrwaechter Platz 1
8011 Grub/Poing

Dr. J. Ladewig
Institut für Tierzucht
und Tierverhalten der FAL
Institutsteil Trenthorst/Wulmenau
2061 Westerau

Prof. Dr. J. Langholz
Institut für Tierzucht
und Haustiergenetik
Universität Göttingen
Albrecht-Thaer-Weg 1
3400 Göttingen

Dr. K. Riemenschneider
Bundesforschungsanstalt für
Landwirtschaft
Braunschweig-Völkenrode (FAL)
Bundesallee 50
3300 Braunschweig

Prof. Dr. A. Rojahn
Bundesministerium für Ernährung,
Landwirtschaft und Forsten
Postfach 14 02 70 - Ref. 322
5300 Bonn 1

Dr. M.C. Schlichting
Institut für Tierzucht
und Tierverhalten der FAL
Institutsteil Trenthost/Wulmenau
2061 Westerau

Prof. Dr. Dr. D. Smidt
Institut für Tierzucht
und Tierverhalten der FAL
Mariensee
3057 Neustadt 1

Dr. H.H. Thielscher
Institut für Tierzucht
und Tierverhalten der FAL
Institutsteil Trenthorst/Wulmenau
2061 Westerau

Prof. Dr. J. Unshelm
Institut für Tierzucht
und Tierhaltung
Universität Kiel
Olshausenstr. 40-60
2300 Kiel 1

Prof. Dr. Rose-Marie Wegner
Institut für Kleintierzucht
der FAL
Dörnbergstr. 25-27
3100 Celle

Dr. K. Zeeb
Tierhygienisches Institut
Am Moosweiher 2
7800 Freiburg

FRANCE

Dr. R. Dantzer
Laboratoire de Neurobiologie
des Comportements
Université de Bordeaux II
146, Rue Léo-Saignat
33076 Bordeaux Cedex

Dr. Fauré
I.N.R.A. Comportement Animal
37380 Nouzilly

Dr. P. Mormède
Laboratoire de Neurobiologie
des Comportements
Université de Bordeaux II
146, Rue Léo-Saignat
33076 Bordeaux Cedex

Prof. Dr. J.P. Signoret
I.N.R.A. Comportement Animal
37380 Nouzilly

Dr. J.P. Tillon
Services Vétérinaires
Station de Pathologie Porcine
B.P. no. 9
22440 Ploufragen

GREECE

Dr. G. Zafiriou
Ministry of Agriculture
Aharnon Street
Athens

IRELAND

Prof. Dr. J. Hannan
Dept. of Preventative Medicine
and Food Hygiene
Veterinary College of Ireland
Ballsbridge
Dublin 4

Dr. O'Connor
Deputy Director of Veterinary
Services
Dept. of Agriculture
Kildare Street
Dublin 2

Dr. P.V. Tarrant
The Agricultural Institute
Dunsinea Research Centre
Castleknock
CO Dublin

ITALY

Dr. Marina Verga
Istituto di Zootecnica Veterinaria
Via Celoria 10
20133 Milano

Prof. Dr. M. Zanforlin
Istituto di Psichologia
Piazza Capitaniato 3
35100 Padova

LUXEMBOURG

Dr. E. Wagner
Administration des Services
Techniques de l'Agriculture
15, Route d'Esch
Boite Postale 1904
Luxembourg

NETHERLANDS

Dr. G. Beuving
Spelderholt Institut
für Geflügelforschung
Beekbergen

Dr. H.J. Blokhuis
Spelderholt Institut
für Geflügelforschung
Beekbergen

Dr. R.G. Buré
Institut für Landtechnik,
Arbeit und Gebäude
Postfach 43
6700 AA Wageningen

Dr. R. de Koning
Research Institute for
Animal Husbandry "Schoonoord"
3700 AM Zeist

Dr. G. van Putten
Institut voor Veeteeltkundig
Onderzoek "Schoonoord"
Postbus 501
3700 AM Zeist

Dr. W. Sybesma
Institut voor Veeteeltkundig
Onderzoek "Schoonoord"
Postbus 501
3700 AM Zeist

Prof. Dr. P.R. Wiepkema
Dept. of Animal Husbandry,
Section Ethology
Marijkeweg 40
6709 PG Wageningen

SWEDEN

Dr. R. Tauson
Swedish University of Agriculture
Science
Dept. of Animal Husbandry
Poultry Division
Funbo-Lövsta
75590 Uppsala

UNITED KINGDOM

Dr. B.A. Baldwin
Institute of Physiology
Babraham
Cambridge

Dr. D.M. Broom
Department of Zoology
University of Reading
Reading RG6 2AT

Miss M.Cherry
Byeways Hook Norton
Banbury OX15 5LG
Oxford

Dr. I.J.H. Duncan
ARC Poultry Research Centre
Roslin
Midlothan EH25 9PS

Dr. R. Ewbank
Universities Federation for
Animal Welfare
8 Hamilton Close
South Mimms
Potters Bar 58202
Herts, EN6 3QD

Dr. R. Moss
Ministry of Agriculture, Fisheries
and Food
Hook Rise South
Tolworth
Surbiton, Surrey

Dr. D.F. Sharman
Institute of Physiology
Babraham
Cambridge

CEC

Dr. J. Connell